THE GREAT WHITE BEAR

The Great
WHITE BEAR

A Natural and Unnatural History of the Polar Bear

Kieran Mulvaney

Houghton Mifflin Harcourt
Boston   New York
2011

For information about permission to reproduce selections from this book,
write to Permissions, Houghton Mifflin Harcourt Publishing Company,
215 Park Avenue South, New York, New York 10003.

www.hmhbooks.com

Library of Congress Cataloging-in-Publication Data
Mulvaney, Kieran.
The great white bear : a natural and unnatural history
of the polar bear / Kieran Mulvaney.
p. cm.
Includes bibliographical references.
ISBN 978-0-547-15242-4
1. Polar bear. I. Title.
QL737.C27M85 2011
599.786—dc22   2010017206

Book design by Melissa Gruntkosky

Printed in the United States of America

DOC 10 9 8 7 6 5 4 3 2 1

To my brothers:
Michael Mulvaney
and
Stephen Mulvaney

And to the memory of my parents:
Peter Mulvaney
(May 19, 1929–October 7, 2008)
and
Wendy Mulvaney
(March 7, 1929–September 10, 2010)

Contents

Acknowledgments

Although mine is the only name on the cover of this book, any undertaking such as this requires the help, encouragement, and support of a multitude of people.

I would like first of all to thank my agent, John Thornton of the Spieler Agency, who quietly and patiently nurtured my proposal and steered it to a good home. I could not have asked him to find a better home than Houghton Mifflin Harcourt; Lisa White immediately showed enthusiasm for, and an understanding of, the project and has been a sensitive and collaborative editor. I very much hope we shall work together on many more titles in the future.

I do not remember how I heard of Robert Buchanan, president of Polar Bears International, or what first moved me to contact him, but it is to my considerable benefit that I did. Robert is not only well connected and highly energetic; he is a genuinely kind and helpful man. It was because of him that I was able to spend time in Churchill, Manitoba, and it was through him that I encountered many of the researchers who have been so helpful in improving my understanding of polar bears. This book would not have been possible without him.

One of Robert's first acts was to put me in contact with the good folks at Frontiers North Adventures; many thanks to John Gunter, Lynda Gunter, and Heather Ross for their assistance before, during, and after my time in Churchill. My time at the Tundra Buggy Lodge and on Tundra Buggies was greatly enhanced by the company, expertise, and professionalism of David Allcorn, Bree Golden, Chris Hendrickson, Trevor Lescard, and Julie Seaton. On Buggy One, I enjoyed and benefited from the companionship and knowledge of Robert Buchanan, B. J. Kirschhoffer, Leann Myers, and Krista Wright of Polar Bears International; Don Moore of the Smithsonian National Zoo; David Shepherdson of Oregon Zoo; JoAnne Simerson of San Diego Zoo; Thomas Smith of Brigham Young University; and Geoff York of the World Wildlife Fund.

I was extremely fortunate that when searching for accommodations in Churchill, I came across a bed-and-breakfast called Duncan's Den; anyone who offers Bailey's to guests with their 6:00 a.m. coffee is all right by me, and I am pleased and honored to be able to call Lance and Irene Duncan my friends. My time with Lance and Irene was considerably enhanced by the concurrent stay of Paul Easthope and Helen Worth, for whom Churchill was one stop on a great around-the-world adventure. During the long train journey north from Winnipeg, I enjoyed enlightening and intelligent conversation with Doug Ross, and I am grateful also for those in Churchill who gladly gave of their time: Tony Bembridge and Jon Talon of Hudson Bay Helicopters; Shaun Bobier of Manitoba Conservation; Mike Spence, Churchill mayor; and Louella McPherson, Don Walkoski, and Marilyn Walkoski of Great White Bear Tours.

My research was considerably aided by those who kindly gave of their time over the phone or in person, including Steve Amstrup of the United States Geological Survey, whose monograph on polar bears made my work a hundred times easier; Robert Buchanan; Richard Harington, Canadian Museum of Nature; Steve Herrero, University of Calgary; Brendan Kelly, University of Alaska; Eric Larsen; David Lavigne, International Fund for Animal Welfare; Don Moore; JoAnne Simerson; Len Smith; Tom Smith; and Geoff York. Brendan, David, Geoff, Tom, and Bruce McKay of SeaWeb all generously agreed to read portions of the draft manuscript. I am tremendously grateful to them for doing so, although I must underline that any remaining errors are mine alone.

For providing photographs, many thanks to Robert and Carolyn Buchanan, Nick Cobbing, Brendan Kelly, and Jill Mangum.

Portions of the Churchill chapter were first published in the *Washington Post Magazine;* thanks to David Rowell for helping me focus my thoughts and sharpen my writing. I am fortunate to be granted a platform to write about polar bears, climate change, the Arctic, and many other environmental issues for Discovery Channel News; my considerable gratitude to Lori Cuthbert, Larry O'Hanlon, and Michael Reilly at Discovery for their support.

Much of my practical experience with the Arctic derives from the two expeditions involving the Greenpeace ship *Arctic Sunrise* that act as bookends to the text that follows. My participation in the first, in 1998, was as a freelance journalist, writing primarily for *BBC Wildlife* and *New Scientist* as well as Discovery News; my involvement was the brainchild of Kalee Kreider and Steve Sawyer, both then with Greenpeace and now respectively with the Office of the Honorable Al and Mrs. Tipper Gore and the Global Wind Energy Council. I am forever thankful to them for launching my life in that direction. During that first trip, I learned much about Arctic wildlife and science from the researchers on board, particularly George Divoky and Brendan Kelly of the University of Alaska; eleven years later, my connection with the *Arctic Sunrise* was as onshore coordinator for an expedition to the eastern Arctic, and although many researchers spent time on the ship during the three months or so that the voyage was under way, I personally came into closest contact with, and learned most from, Ruth

Curry, Jim Ryder, and Fiamma Straneo of Woods Hole Oceanographic Institution; Gordon Hamilton of the University of Maine; and Leigh Stearns of the University of Kansas. Gratitude and greetings to those at Greenpeace who helped make either or both of those expeditions come together, particularly Willem Beekman, John Bowler, Melanie Duchin, Thomas Henningsen, Beth Herzfeld, Paula Huckleberry, Frank Kamp, Martin Lloyd, Sharon Mealy, Dan Ritzman, Sallie Schullinger-Krause, Adam Shore, Walt Simpson, Arne Sørensen, Dave Walsh, Pete Willcox, and the many others who worked tirelessly on board or ashore.

As I have commented in acknowledgments in earlier books, authorship is a mostly solitary and oftentimes lonely process, made palatable by the support of friends across the country and around the world. I think particularly of Rachel Charles, Melanie Duchin, Mike Harold, D. J. and Catherine LaChapelle, Stephanie Lasure, Loni Laurent, Bruce McKay, Pilar Vergara, Sam Walton, Jim Weber (who, with Jason Colosky, Terry Hansen, and Scott Hansen, manages to get me out of the house and into a nice restaurant once every couple of months or so), and especially Kennedy Clark.

I mean to imply no disrespect at all to those I have already mentioned—nor, I suspect, will any be inferred—when I say that, most of all, I cherish the love of my family, to whom this book is dedicated: my brothers, Michael and Stephen; my darling mother; and my late father.

When I lived in Alaska, Dad particularly enjoyed chuckling at the notion that I would find a polar bear outside my door one day. I always assumed he was joking, but it was often not easy to tell with my father, whose eyes displayed a near-permanent twinkle of good-natured mischief. He would have loved seeing this endeavor come to fruition, and it was always my intention to dedicate it to him; but shortly after I started work on this project, he began his final journey. Much as I have enjoyed, and take pride in, this book, the memory of its creation will always be bittersweet and forever entwined with the images of those final few months, as I alternated between typing on my computer in my childhood bedroom and, with my brothers, sitting at my father's bedside.

I love you and miss you, Dad. May you experience an eternity of fair winds and following seas. You have earned no less.

Kieran Mulvaney
Alexandria, Virginia

Journey

The bear was near the horizon when we first saw it.

A dot in the water, barely visible above the waves, it was not, initially, obviously a bear at all.

"Look at the size of that seal," exclaimed the mate, raising binoculars and prompting the captain to do likewise.

There was a pause as the two men pondered the distant object, perhaps realized what they were looking at, dismissed the thought, returned to it, and finally conceded what was increasingly clear.

That a polar bear should be in the vicinity should not, on the face of it, have been particularly remarkable. We were, after all, anchored just off the north coast of Alaska; however one defines the Arctic— and scientists, geographers, and oceanographers debate many conflicting and complementary delineations—we were undoubtedly in the heart of it and deep within the polar bear's realm.

Yet the initial confusion was understandable. Polar bears are creatures of the ice; but, save a few floes drifting past in the current of the Beaufort Sea, there was almost none to be seen—just mile upon mile of open water.

We had come in search of the edge of the Arctic Ocean sea ice. The boundary where open water progressively yields to its frozen counterpart is an oasis of marine life, one that our passengers, biologists from the University of Alaska, were keen to reach. But the ice edge had retreated to the north, earlier and farther than normal; it would take us many days of steaming to reach our goal. There was no way of knowing how long or how far this particular bear had been swimming, but its chances of ever finding its species' preferred habitat were all but nonexistent.

It was a Sunday morning. The scent of freshly baked bread and of the breakfast that was cooking in the galley wafted from deck to deck and into the crisp arctic air. It filled our nostrils as we tumbled from mess room and cabins, hastily pulling on fleeces and coats, to watch as our visitor approached. The aromas stretched far beyond our green hull, wafting into the distance, their decreasing strength more than compensated for by the extra sensitivity to them on the part of the bear—which, it was increasingly clear, was not simply swimming in our direction but making a determined beeline for us.

It paddled closer, close enough that now we could see it clearly, its paws working feverishly beneath the surface of the water, its long neck straining to keep its head above the surface, its eyes fixed eagerly on the steel grail ahead of it, its small ears flat against the side of its head. A passing ice floe provided welcome respite and the bear took advantage, clambering out of the ocean, its fur thick with water. It shook itself briefly, walked from one end of the floe to the other to stay level with the ship as the ice drifted past, then plunged back into the water and paddled closer to us once more. Another floe arrived, and again the bear climbed upon it, rested there until it began to drift out of range, reentered the water, and swam toward us again.

Two or three times it repeated the process, each occasion appearing to be progressively more taxing as the bear fought to drag its waterlogged weight onto the ice, its shoulders seeming to sag ever so slightly with each repetition and the growing realization that any hope it might have had of clambering on board was destined not to be realized.

Eventually, it gave up. Having hauled itself onto a passing floe for perhaps the third or fourth time, it chose not to subject itself anymore to the rigors of swimming in the Beaufort Sea on a hapless quest. Its mouth open, it tore away its gaze, looking alternately down at the ice beneath its feet and into the distance, anywhere, it seemed, except directly at the object of its desire and frustration. And we watched as it stood there, forlorn and defeated, drifting into the distance.

For those of us assembled on deck, it had been a thrilling diversion. We had alternately gasped in awe and made the kind of cooing noises normally reserved for watching a puppy do tricks or a baby crawl along the floor. The bear was cute and furry and, from our vantage point on deck, perfectly harmless. Its apparent attempts to find a way on board had been endearing, precisely because they had failed.

The bear did not, of course, share with us its motivation, but it seemed clear enough that what, for us, had been diverting and entertaining had been, for our visitor, entirely more serious and desperate. The extent of its determination and focus was underlined by the fact that it had been swimming into the current — a current that, a few days earlier, had demonstrated its strength by driving an enormous floe into us with such force that it had dragged our anchor from the sea floor and pushed our ship helplessly along the coast for over an hour until the captain extricated us. As we looked down on that bear, it focused keenly on the smell of food — the bacon, the bread, the strange bipedal seals waddling around on deck — that lay tantalizingly out of reach.

I find it mildly astonishing in hindsight, but I had no comprehension at the time of just how unhealthy that bear clearly was. The fur on its nose was patchy and brown. By polar bear standards, it was painfully thin; as I look years later at footage we shot on that day, I can clearly see the bear shivering in the water, a function of having lost so much of its insulating body fat.

But I had at the time no frame of reference. The only other polar bear I had ever seen was a captive in the local zoo I sometimes visited as a child. Confined in an enclosure that provided neither the space nor the stimulation required by an intelligent, inquisitive mammal hard-wired to wander for vast distances across the pack ice, it paced continuously back and forth—two steps forward, a sideways shake of the head, two steps back—minute after minute, hour after hour, day after day. For the bear that had swum up to us that morning in the Beaufort Sea, its torment was not psychological like that captive but physical: unable to reach its favored habitat, the great predator of the Arctic was forced to struggle for scraps of food and, judging by its condition, was faring poorly in its quest. It seems in retrospect highly unlikely that it lived for very much longer after it drifted away from us.

Polarbjørn is both a Norwegian word for "polar bear" and, appropriately, the original name given to the target of that particular bear's attention: a green-hulled, 150-foot-long icebreaker now called the *Arctic Sunrise*. Built in Norway in 1975, the ship's initial *raison d'être*, like that of its mammalian namesake, was to hunt seals: an ironic start to life for a vessel that was now used as a platform to document environmental damage and highlight the impacts of climate change. That change of vocation had taken effect in 1995; two years later, the *Sunrise* became the first ship to circumnavigate James Ross Island in the Antarctic Peninsula—a feat that had been impossible until warming temperatures led to the disintegration of part of the Larsen Ice Shelf, which had previously anchored the island to the Antarctic mainland. Shortly afterward, it traveled to the other end of the world, grinding through Arctic ice in the Beaufort Sea, which is where, one year later, we were gazing down on our unexpected visitor.

There was, however, one slight problem.

We were on an icebreaker, but there was very little ice. And as the ice eluded us, so too did the ice denizens we had come to see.

The only evidence that polar bears were anywhere in the neighborhood came from a stopover at Deadhorse, the supply post for the oil fields at Prudhoe Bay. On the inside of the door to the town's general store, a sign warned visitors to exit carefully. "There are bears outside," it warned. "Look before leaving. Bears are soooo cute, but their claws and teeth are sharp and they like to maul people." On the outside of the door, there was a freshly pinned notice, advising that a polar bear had been seen around town that very morning and that extra attentiveness was therefore warranted.

Two days later came the unexpected visit from our swimming, skinny bear; but after it had drifted forlornly into the distance, we saw no further sign of ice-loving ursids — or, indeed, much sign of ice at all — in the immediate area.

We steamed west and north, away from the Alaska coast and toward the Russian Arctic, in search of an ice edge. Within a week, we had found it, and when we did so, it came upon us neither gradually nor incrementally but with a suddenness and emphasis that threatened to grind us to a halt.

The *Arctic Sunrise*, while an icebreaker, was a relatively small one; it possessed nothing like the heft of, for example, the nuclear-powered Russian behemoths that plow their way to the North Pole each year for the benefit of paying tourists. Like all icebreakers, however, its hull was rounded and without a keel; that made the *Sunrise* an uncomfortable ride on the open ocean, as it bobbed around more like a cork than a seagoing vessel, but it enabled it to rise onto ice flocs with relative ease, a characteristic of which Captain Arne Sørensen took ready advantage.

Smaller floes he would simply nudge to one side or thump out of the way; with larger, more formidable adversaries, Sørensen demonstrated an entirely more sophisticated and subtle technique. He would head directly for one, slow down shortly before reaching it and perhaps even put the engine into reverse, relying on the ship's momentum to carry it forward, and ride the ship gently onto the ice. Then once more he would crank up the engine, powering the ship forward and pushing the floe out of the way, in the process frequently steering the ship hard to starboard or port so that as the ice began to move, the vessel would fall away into open water. As it did that, Arne would ease up on the throttle anew to prevent the ship, suddenly no longer meeting resistance from a huge chunk of ice, from rushing headlong into another.

Initially, the ice edge encounter brought a rush of excitement to the *Arctic Sunrise*; crew interrupted watches and deck duties to lean over the bow and watch as Sørensen nudged small floes out of the way and slowly ground his way over and past larger ones. But as the afternoon unfolded, the ice became progressively thicker, and the bridge grew quieter as brows furrowed and focus sharpened. The percentage of water that was covered with ice now exceeded the percentage of water that wasn't, and ocean currents pushed floes into one another, compressing them, closing the ice into a mass that accumulated astern of us. Sørensen ascended to the crow's nest, affording him not only greater visibility but a separate set of steering controls, enabling him to look for leads through the ice and direct the ship toward them. The ship was enveloped in quiet, the scientific team peering keenly through binoculars in the hope of spotting marine life, the crew alternating between enthusiasm and uncertainty, the only noises the gentle throb of the ship's engines followed by a crash as the bow collided with a floe and the hull shook in protest at the force of the impact.

The task was not aided by the evening descent of fog, combining with the day's diminishing daylight to restrict visibility at times to no more than thirty yards. The ship slowed to a crawl as pieces of ice loomed out of the mist like ghouls on a fairground ghost train. When finally the fog lifted, the sense of relief was palpable.

"It is," observed the phlegmatic Sørensen, "so much easier when one can see where one is going."

When the fog had lifted, the view before us was as different as it was now starkly beautiful, as if in its ascent the mist had peeled away the largest, most threatening floes, leaving in their place a dead calm sea, with nary a ripple, let alone any kind of swell to disturb its flat surface. Scattered about were innumerable smaller chunks of ice, many of them curiously misshapen and twisted, all of them now drifting harmlessly past. It looked for all the world like the detritus from an unseen tornado, a scene of terrible yet beautiful devastation, pieces of ice strewn randomly across the sea surface by the hand of God.

We continued onward, north and west. Alaska receded far into the distance as we steamed into the waters of the Russian Arctic and closed in on our ultimate goal, the twin sentinels of Wrangel and Herald islands. Now we were certain we would find what we were looking for. Here, a combination of high latitude and the vagaries of ocean currents ensured that, even in late summer, sea ice was plentiful in thickness and extent. As a result, Wrangel and Herald combined to create a kind of polar bear paradise: a place where even under the harshest of conditions, the food supply was relatively plentiful and stable.

We steamed slowly along, poking our way through floes, grinding along the edge of the fast ice that remained firmly attached to the islands' coasts and stretched for miles outward, as the islands themselves reached defiantly upward before disappearing into the fog. Dark clouds of sea birds swarmed through the skies. As the hull of our icebreaker rose up onto, and then cracked through, the ice in our path, schools of tiny arctic cod scattered rapidly, desperately searching

for shelter. Here and there we spied ringed seals. Ahead of us, looming out of the mist, an occasional floe revealed itself to be packed
tightly with walruses, seeking safety in numbers from the predator that
always prowled nearby.

And what a predator.

Our crew was a hardened bunch, accustomed to being at sea for
weeks or months at a time, used to seeing wildlife and landscapes
known to most people only through television. But when, after weeks
of slow-building anticipation and days of fighting through the ice, the
call finally came from the bridge that a polar bear was up ahead, it
prompted the instant dropping of tools or abandonment of food in a
mad rush to the bow.

This was a polar bear the way polar bears were meant to look: its
fur thick and plush, its rump healthily rotund, its shoulders impres-
sively muscular. It seemed not just to be walking across the ice but
swaggering, as if surveying its kingdom and daring anyone or anything to encroach upon it. It carried, as only the most dominant predators do, the confident, almost arrogant bearing of an animal that felt
no threat from fellow denizens in its kingdom. Its massive shoulders
rolled with a self-assured poise reminiscent of an undefeated prize-
fighter.

"Oh, yeah," said one of the crew admiringly as we hunkered down
to shelter from a biting wind that had no evident effect whatsoever on
our companion. "Look at him. He owns the ice."

The bear seemed at first unaware of our proximity; but then it shuffled to the edge of the floe it was patrolling, sniffed the air, looked in
our direction, briefly appeared to brace itself as if to jump into the
water and swim toward us. It changed its mind, its ignorance of our
presence evidently yielding to indifference toward it, turned around,
and resumed its wandering, scratching at the ice, occasionally looking
in our direction but mostly disregarding us as it continued on its way.

That day, and in days subsequent, there were several more encounters, several more occasions on which the *Sunrise* cut back its engines and eased quietly along an ice floe as a bear sauntered confidently nearby, several more instances when work would halt as, transfixed, crew and scientists alike stood silently as they watched the scene unfold in front of them.

At times it was hard not to imagine that greater forces of Nature were working to emphasize the rarity of the experience and the majesty of the animal at the heart of it. As if an unseen stagehand were operating a celestial spotlight, the sun abruptly shone on one bear just as we approached the floe that was its realm. Playing to the audience, the bear yawned theatrically and lay down, tucked its paws under its chin, placed its snout on them, and closed its eyes—occasionally opening them just a fraction to keep watch on the strange green iceberg in its midst. Then, receiving its cue from a director in the wings, it hauled itself to its feet, walked into a hollow in the ice, and posed for pictures with its front paws resting on the cavity wall.

Most regarded our intrusion with mild curiosity or indifference. Others appeared resentful or even hostile. One marched away, looking repeatedly over its shoulder like a reproached dog. Another, uncertain what to make of the interloper yet unwilling to back down in the face of its enormity, stood its ground, bobbing its head and hissing at us in a combination of fear and ferocity.

But then, unexpectedly, a bear would reveal a more vulnerable side. We slowed down, drifted to a halt, nosing into the ice as a bear lifted its head, sniffing the air and scenting our proximity. Slowly—not cautiously, but unhurried—it ambled toward us, curious as to the nature of the large object that had suddenly invaded its terrain. Above, seemingly unnoticed, a crew of thirty people leaned over the starboard bow, holding their collective breath and daring not to make a sound as the bear walked up to the hull, stretched its neck, sniffed, and all but touched its nose to the metal.

Then there was a high-pitched beep and the click of a camera shutter, and at once the bear whipped away from us, racing back across the ice before slowing down, looking over its shoulder, and then continuing in a lumbering trot. It paused, looked back toward us now, sniffed the air again, and as if to reassure itself that a polar bear could not possibly truly have been frightened by anything so inconsequential, seemed almost to square its shoulders as it recovered its natural swagger.

It strode up a small hummock, looked briefly in our direction once more, then disappeared down the other side and was gone.

Becoming

In the dark, beneath the snow and ice, the cubs stir.

It is almost spring in the Arctic, and the cubs are three months old. But they have yet to experience the warming glow of sunlight or the chill of the polar wind. The world in which they have spent the entirety of their short lives is a den hollowed out of snow, barely large enough to contain the cubs and the mother against whom they are curled tightly.

She had been inseminated in May, fully ten months previously, a few weeks after she had entered heat and attracted the attention of most of the males in her vicinity. But insemination did not lead at once to pregnancy. The eggs, although fertilized, did not immediately implant.

Once the cubs were born, the female would not eat for at least four months. So the eggs were held in abeyance while she readied herself for the onerous task that lay ahead; and in preparation for that task, she ate. More accurately, she gorged.

In this she was aided by the forces of evolution, which had ensured that the time in which she would be seeking sustenance was a bountiful one, for spring in the Arctic is a veritable polar bear buffet. This is the period when the seals they eat are both most abundant and most vulnerable to predation, when ringed seal pups are born and, six weeks later, weaned. Upon weaning, each pup may weigh up to sixty pounds, forty-five pounds of which might be calorie-laden body fat. And, at this young age, not one of these corpulent morsels has learned to fear or avoid predators, leading to what noted polar bear researcher Ian Stirling has described as a "superabundance of fat, naïve seal pups that enables the pregnant females to accumulate fat so quickly."

They need to. Later in the summer, the sea ice breaks up, the seal pups disperse into the water, and the smorgasbord is over. In some parts of the polar bear's range, for example, north of Russia's Wrangel Island, the feast may persist a while longer, the females continuing to hunt on the drifting pack ice even as it begins to break apart and drift south, until eventually the floes grind against the shore and the bears come onto land. Even then the respite may be brief, the bears needing to shelter only a few weeks before returning to the hunt when the sea freezes anew.

At the other extreme, in Canada's Hudson Bay, the sea ice melts entirely by the end of July, forcing all the bears ashore. Because the ice does not form again until early November, they have no opportunity to hunt more seals before giving birth. As a consequence, a pregnant female in Hudson Bay may go eight months without any nourishment at all, almost certainly the longest period of food deprivation of any mammal on Earth.

Under such circumstances, the obvious solution might seem to be for the eggs to implant, and the cubs to be born, as soon as the bears are forced to leave the ice. But however much (were they able to dwell on such matters) the pregnant females might wish it could be so, they must wait.

They must wait to give birth so that the cubs are weaned in spring and not the depths of winter. And they must wait until the weather

once more cools and the falling snow forms drifts large enough for
them to fashion dens.

The precise timing at which they do so varies according to geographical location and the vagaries of each year's weather conditions,
but its approximate schedule remains relatively constant, following the
path of winter's onset from north to south. In the Canadian Arctic,
bears enter their dens on average by mid-September; in Alaska and
northern parts of Svalbard (an archipelago north of Norway), they
generally do so by mid-October; in more southerly parts of Svalbard,
it is not until late October or early November.

Not only pregnant females take shelter in dens, although they alone
do so for the duration of the winter. Many other bears also seek ref-
uge to avoid storms and extreme cold or heat, or when hunting is
poor, particularly in the period after the sea ice breaks up and before it
re-forms. It is a phenomenon seen less often in the High Arctic, where
sea ice is generally available year-round, but observed with particular
frequency at the southernmost reaches of the species' range, in Hudson Bay.

There, when the sea ice melts in summer, bears come ashore en
masse and hole up in earthen dens, where the temperature is cooler
and bothersome insects are less likely to intrude. Most of these dens
were excavated in eons past by ancestral bears; over time they have
grown, and continue to do so, as a result of the bears' body heat gently melting the permafrost beneath the surface.

Following the onset of fall and the return of the sea ice, males, females already with cubs, and females for which maternity is only a
memory return to the hunt. Pregnant females that have also taken advantage of the topography, however, tarry a while, waiting until the
drifting snow has covered their shelter to sufficient depth that they can
tunnel into the drift and dig out a new den for the winter.

At its most fundamental, construction of a maternity den requires drifted snow in which to dig a hole, and still-drifting snow that

will cover up the hole and the bear that has curled up inside it. The first consideration for any pregnant female, says Thomas Smith of Brigham Young University (perhaps the foremost authority on denning behavior), is security: selecting a spot that minimizes the likelihood of exposure to or interactions with other bears. Cannibalism has been recorded in all three bear species in North America—Smith and colleagues have recorded it occasionally in polar bears—and a nursing female with cubs is potentially vulnerable to any kind of attack. Accordingly, whether they den on pack ice or on land, females invariably seek to avoid areas where males are actively hunting. But in addition to that one fundamental concern, there are subtleties and nuances that make some locales more preferable denning sites than others.

Richard Harington, a polar bear biologist formerly with the Canadian Wildlife Service, found that in the Canadian Arctic, the vast majority of maternal dens face south. Prevailing winds from the north deposit greater amounts of drift, and, as with human housing, south-facing real estate is valuable in the polar bear world for its extra exposure to warming solar radiation. On Wrangel Island, in contrast, the distribution of dens follows no such discernible pattern; highly variable winds cause snow to build up on all sides of the island, making most areas suitable for denning.

Indeed, Wrangel Island's sheltered, mountainous landscape provides ideal denning habitat. Other locations where drifts are easily formed—where there is an abundance of earth banks or hillsides, where there are valleys or mountains—boast a higher concentration of dens. Elsewhere, however, suitable spots are harder to find.

Along the North Slope of Alaska, the terrain is flat; snow is blown by unforgiving winds across the tundra, and bears must make use of drifts wherever they can find them. In many cases, dens are hollowed out of drifts as little as four feet high, nestled along coastal and river banks.

Most dens are near the coast. In the Canadian Arctic, Harington found that 61 percent of dens were within five miles, and 81 percent within eight miles, of the shore, and a 1985 study found that most dens on Svalbard were no farther inland than two miles or so. The pattern is broken only in Hudson Bay, where the coastal plain is boggy and flat, and where females must sometimes trudge over sixty miles to find suitable terrain.

The dens' proximity to shore provides easier access to the seals the desperately hungry mother will need to replenish herself when she and the cubs emerge in the spring. But some bears take this a step further and actually make their dens on the ice itself: either land-fast ice, which is attached to the shore, or even the drifting pack. A 1994 study of Alaska bears fitted with radio collars found that over half the dens those bears made were on sea ice—a high degree of ice denning that is not found anywhere else. It might be a consequence of the flat terrain of northern Alaska making suitable sites on land relatively scarce, but Steven Amstrup of the United States Geological Survey (USGS), who has likely spent more time studying the polar bears of Alaska than anyone else, believes that the sea ice in the southern Beaufort provides a more solid foundation than is the case elsewhere.

"Historically, the Beaufort Gyre circulated multiyear sea ice through the Beaufort Sea," he says. "That's ice that circulated through the Arctic year after year and kept getting thicker and thicker, so the sea ice in the southern Beaufort Sea had some of the most stable sea ice anywhere in the Arctic. It probably provided a very stable platform for polar bears to den on."

Researchers elsewhere have concluded that denning on sea ice is infrequent and often limited to bears that could not make it ashore before the ice broke up. Understandably so, for making dens on the pack poses risks not experienced by those that choose to den on land. Ice floes shift, break up, re-form; they can turn over or raft under other floes with which they collide. The authors of the aforementioned 1994 study, among them Amstrup, saw six polar bears in pack-ice dens that were swept past Point Barrow, the northern tip of Alaska, and southwest into the Chukchi Sea because of unusually unstable ice conditions. The dens of two of the females had been destroyed when ice floes had collided and rafted onto each other; in their mouths the females carried tiny cubs. The researchers saw the mothers again later that spring; of the cubs, however, there was no sign.

A pregnant bear knows instinctively and precisely what to look for in a denning site. She walks from snowbank to snowbank, testing each for consistency and depth, poking and prodding until she finds one that meets her criteria. She may travel many miles and take several days until she is satisfied with a location. Then, when she has found a spot she likes, she digs.

She will first excavate an entrance tunnel, which is normally about six feet long but so narrow — rarely more than two feet in diameter — that it is a wonder a fully grown, pregnant polar bear with a winter's worth of fat reserves could possibly squeeze through. Then she will angle slightly upward, so that the warm air does not escape through the passageway, and create the main chamber in which she and her cubs will spend the next several months. Some females, for unknown reasons, become creative, adding an antechamber or even two — although in many cases the most extensive additions are made after mother and cubs have emerged from the den but before they have opted to leave it behind completely. Thomas Smith has found a few dens with "interconnecting tunnels up to thirty-five feet in length, with as many as four entrances." There may or may not be a small ventilation shaft.* The main chamber will itself measure approxi-

* Smith reports that of all the dens he has studied on Alaska's North Slope, "there has been no ventilation shaft . . . none. How they can breathe in there is truly puzzling. I'm convinced a human in there would die. So how do they do it?"

mately six and a half feet by five feet, and about three feet high: little room for the young family to do anything except remain curled up together.

This will be the mother's home for the next four months. When the chamber is completed and the entrance shaft has been resealed with drifting snow, she lies down and prepares to give birth.

Throughout the polar bear's range, there are approximately fifteen areas where dens are most concentrated, where geographical convenience and topographical advantages combine to create ideal conditions—specifically, a plenitude of sheltered areas with drifted snow—for den formation. Between them, the Russian Arctic islands of Wrangel and Herald host between 350 and 500 pregnant polar bears each year, approximately 85 percent of the breeding females in the Chukchi Sea population. In places, twenty dens may be squeezed into an area of a little more than a square mile.

In August 1990, biologist Nikita Ovsyanikov arrived on Wrangel to begin a detailed field study of the islands' bears. He set up base at a point called Cape Blossom, where bears congregated in search of food, in the process passing within yards of his cabin. The bears at Cape Blossom are mostly males; during Ovsyanikov's study period they proved inquisitive, seemingly fearless, and apparently unconcerned by the researcher's presence.

The bears on Herald—home to such an abundance of maternity dens that the scientist dubbed it the region's polar bear nursery—were different, as indeed was the island itself. During our visit on board the *Arctic Sunrise*, our grizzled bosun sized up the otherworldly vista that thrust straight up from the ice into a low layer of clouds and declared it to be "like something out of Arthur Conan Doyle's *The Lost World*." Ovsyanikov, too, noted Herald's "severity" on his first visit, observing that the "tall, jagged, almost vertical cliffs plunged straight down into the cold, dark green water with no beach to break the fall." It did not have anywhere obviously comparable to Cape Blossom, no areas for the congregation of seals or walruses and thus of polar bears; its attraction was purely the plethora of mountainous but sheltered nooks and crannies on its otherwise inhospitable surface. Wrangel may have been the kingdom of the ice bear, opined Ovsyanikov, but Herald was the castle of the snow queen.

Accordingly, the only bears on the island at the time of the researcher's visit were pregnant females preparing to den, and their behavior, Ovsyanikov observed, was markedly different from that of the bears on Wrangel. Whereas the bears at Cape Blossom would frequently approach the scientist's cabin, those on Herald studiously avoided it, even when he placed a box of thawing reindeer meat outside. These bears, he surmised, were clearly not only very full after a period of gorging but were extremely cautious and sensitive.

Just how cautious and sensitive became clear following an unintentional encounter that proved upsetting to both bear and human, when Ovsyanikov came unexpectedly upon a bear that was digging a den. He did not notice her, he wrote, until he had drawn level with the hollow she was creating; he saw the bear, which had her head down as she dug, before she saw him. He immediately dropped to the ground to avoid detection. But it was too late. The crunching of snow underfoot had given him away.

The bear, he wrote, lifted her head, looked at him with surprise and fear, and "hissed so loudly that I thought she was going to turn herself inside out." She stared at the scientist for several seconds before climbing up a steep slope away from him, still hissing, clearly struggling to move her distended body.

Ovsyanikov, a conscientious researcher, was deeply distressed that, however unwittingly, he had caused the bear to abandon the site she had selected, forcing her to search for a new suitable location. His presence had altered the animal's behavior, a violation of the field biologist's prime directive; as a result, the pregnant female—which,

with resources at a premium, had little margin for error—was forced
to expend extra energy she could ill afford to waste.

Even so, for a bear to desert a den during construction is neither
unprecedented nor unusual. Researchers have found that abandonment of dens in favor of more agreeable sites can happen with some
frequency during the late autumn months when females are looking
for somewhere to bed down for the winter.

When the dens have been dug and the cubs born, however, it is a
different story.

Ian Stirling, who has spent many decades studying the polar bear
population in Hudson Bay, has noted that once pregnant females
have become established in their dens, they seem determined to remain there, to the extent that they appear almost passive in the face
of disturbance.

"I think their main interest is in detaching from the bad weather
outside," offered Richard Harington, who in the 1960s conducted the
pioneering study of polar bear denning behavior. "They want to be
as secure as possible. They go into what is called a carnivore lethargy:
their temperature drops, but it's nothing like hibernation. You can disturb them fairly easily. They can get up and start walking around. I've
taken bits of the roof out over dens, and you can see the female inside. At first you don't hear anything but heavy breathing, and then
when you poke through the roof you sometimes hear a bit of a growl.
When you open it up you can see the female walking round inside,
and sometimes the cubs against the far wall."

For the females, it is of course important to conserve as much energy as possible once bedded down; they will need every ounce they
can find, given that they will be surrendering calories to their cubs
without eating for at least four months. Far better to sit and wait out a
potential threat than to run or fight.

Which is not to say that neither of the latter ever happens.

"I remember one time we were in northern Southampton Island,

and we were just building an igloo for the night," Harington told me, with a chuckle at the memory of the tale that was to come.

"I was with Tam Eeolik [Harington's Inuk guide] and [Eeolik's son] Tony. I took one of the bear dogs; I could see some ruffled snow on the edge of a rise. So I told Tony that I was going to go up and have a look at it. So I took the best bear dog we had—and these are dogs that are specially trained to go after bears, to corner them, to hold them for the Inuk to come up and kill them. So anyway, I took this husky and went up, and it looked like it was an abandoned den; there was lots and lots of snow below the entrance area. I started digging away at the entrance and the bear dog that was up with me went up onto the snow above me, and all of a sudden the roof fell in and this female came out and reached for the dog, and this famous bear dog vanished over the hills in no time. I could just see the end of his tail."

Harington is not the only one. Geoff York, now a polar bear expert with the World Wildlife Fund but at the time a researcher at USGS, recalls an occasion when he and Steven Amstrup had an unexpectedly and uncomfortably close encounter with a den's inhabitant.

In an attempt to learn more about denning behavior and den construction, USGS researchers had attached radio collars to a number of pregnant females the previous fall and, with the arrival of spring and those females' departure (as determined by the collars' GPS readings) for the sea ice, York and Amstrup visited the now-empty dens, into which they crawled and which they measured and documented.

"In the course of that, we stumbled across some dens of non-collared bears," he recalls, "and we were always particularly careful in those instances because we didn't know if anybody was at home or not." One of those dens was between two that had been built and occupied by bears that wore collars and, with their cubs, had departed; all three were within a hundred yards of each other. Approaching by helicopter, the biologists surveyed the scene from the air and landed.

"We shouted and hollered, threw snowballs into the den," he continues. "Steve Amstrup stuck his head into a hole and looked around. Nothing. We shut the helicopter down and got the gear out. And polar bear dens are mouse-hole-like — the entrances are much smaller than you or I could comfortably squeeze into, so we typically make them larger. And Steve was about to do that when his leg sort of popped into the tunnel and he heard a hiss. And he just about had time to say, 'Bear' before she came out. And he was standing in her exit hole. Steve went from that hole to the helicopter in nothing flat; I looked up and he was gone. I took a step backward, stepped into a crevice, and fell backward and landed on my back like a turtle."

Like Harington, York chuckles at the memory and makes light of parts of it — a benefit of having survived the incident.

"She saw me on my back and thought, 'Well, if you're going to make it easy . . .'" He smiles. "She starts walking over, and she looked enormous, and the only thought that passed through my head at the time was 'So this is how it ends.' I did have a .44; lucky for me and for the bear, it was in a holster that was completely inadequate and it was difficult to remove. My eyes were fixed on her and as she came closer, I yelled out, 'Steve!' as loudly as I could. I saw him out of the corner of my eye wheel round with his sidearm drawn, and luckily in all the commotion, the pilot saw what was happening, fired up the helicopter, and she took off."

That the bear took so long to respond to the scientists' presence highlights the fact that, particularly once cubs are born, it is clearly more desirable to hunker down than to take action that might expose the young family to the winter elements. The fate, described earlier, that befell the cubs whose mother had denned on the pack ice may have been a consequence of the riskiness of that particular environment; but the odds of survival of cubs exposed to the winter air are no greater for cubs whose den had been on land.

Furthermore, it seems fair to infer that mother and cubs feel secure in their little home. For example, field studies have shown that denned polar bears seem remarkably tolerant of nearby human activity, including aerial and ground traffic; only when helicopters took off or landed within a matter of feet of the den did researchers even record the bears making any kind of noise in response. Buried in their snowdrift, together in their warm, dark sanctuary, much of the time they are blissfully unaware of, and largely unconcerned by, any threats that may be passing by in the world outside.

In the den it is warm, the chamber's insulated properties and the body heat of the bears combining to make the air inside as much as 70°F warmer than the Arctic winter outside.

The eggs became implanted once the female found a spot for her den; a few weeks after she had made her home in the snowdrift, the cubs were born. Polar bears sometimes give birth to three cubs, rarely four, almost never one. On most occasions, as in this instance, there are two.

When they were born, seven months after the mother had been impregnated, they were blind and weighed roughly two pounds, but they were already covered with a downy coat of fur—so gossamer-fine that newborns are sometimes inaccurately referred to as "naked" or "hairless"—and their tiny paws were tipped with sharp claws. That latter feature enabled them to clasp more tightly to their mother's coat as they suckled. To our taste buds, the milk they ingested would be rich and rank: the fat content of polar bear milk is on average a little over 30 percent, compared to just 4 percent in human breast milk. But it is rich for a reason: high in protein as well as fat, it enables the cubs to grow rapidly.

At roughly thirty days, their eyes opened. Twenty days later, they sprouted their first teeth and started to develop a sense of smell. Despite the confined space, however, there was little, beyond the body odors of the three bears, for that newfound sense to discern: the mother kept the den spotless, immediately burying the cubs' feces and urine in the snow. (She, consuming no food and imbibing no liquid, expelled no waste of her own.) By the time they were two months old, the cubs began exploring their surroundings and adding to them, carving out their own small chambers as they tumbled with each other and crawled over and around their mother.

Now, a further month later, they are ready. Already the cubs have grown to sixteen pounds in weight, and in providing the nourishment that has enabled her offspring to grow so rapidly, the mother has depleted most of her resources. The cubs are set to make their entry into the outside world, and she is desperate to eat. Arching her back, she powers upward, breaking effortlessly through as much as three feet of hard, wind-packed snow. "A human could *never* break up through that ceiling," marvels Thomas Smith, "but polar bear mothers have no problem."

Having created a hole in the roof of the den, she sticks out her nose and sniffs the air. Sensing no danger, she slowly emerges, and, after confirming that the path is clear, with a soft, low grunt she signals to her cubs that it is safe to join her. With that, they scramble through the opening and for the first time experience the world outside the den.

It sounds on paper the slightest of shelters for the most powerful of predators. A hole in a snowdrift, sealed by more snow, scarcely seems sufficiently substantial to provide privacy and protection for one of the largest truly carnivorous mammals on Earth. And yet, the hostile environment is an impediment to all but the most curious and determined, and the monochrome surroundings render the dens invisible to all but the keenest, most experienced eyes.

"During the months of December and January, there's no evidence of any dens anywhere," says Mike Spence, a Cree tracker, owner of the Wat'chee Lodge near the polar bear denning area south of Hudson Bay, and a guide for photographers and naturalists wanting to see bear dens, and mothers and cubs emerging from them. "It's snow, it's creeks, it's willows, it's all of that. But then in February and March, well, a lot of it is experience, but you're looking for certain things. You know where some of the locations are, you travel, you track, there's been some disturbance so you look for that. If you see tracks, follow the tracks. It's difficult, but you keep at it."

Photographer Thorsten Milse, whose book *Little Polar Bears* is a pictorial paean to the inherent adorableness of furry white bear cubs, willingly admits in his tome that without the expert guidance of Mike's brother Morris, "I would not have been able to take a single shot of a polar bear, nor would I have been able to catch a glimpse of white fur."

Finding the location is just the beginning; while the emergence of females and cubs can be approximately predicted at a given location down to a matter of weeks, greater precision is impossible and there are no warning signs, no indicators of imminent appearance. One moment there is a snowdrift, the next a black nose, curious eyes blinking into the sunlight, a polar bear mother seeing the world anew and her cubs taking it in for the first time. With just the slightest loss of concentration, the briefest aversion of gaze, the moment can easily be missed.

In 1982, filmmakers Hugh Miles and Mike Salisbury endured a tortuous month attempting to film a den opening on Svalbard, for a BBC documentary series entitled *Kingdom of the Ice Bear*. Setting out from the archipelago's capital on March 4, they endured several days of storms, an overturned snowmobile, the extreme discomfort of camping on the sea ice and trying to distinguish between cracking floes and approaching bears and wondering which was worse, and the frustration of two false alarms—setting up to film what turned out to be a temporary den, and recording cubs that on review appeared to be suspiciously large and ultimately proved not to be newborns at all.

On March 29, the filmmakers and their guides stood in the remnants of a temporary snow shelter they had built that had been ruined when, in their absence, it had been visited by a bear that "rather inconsiderately re-emerged straight out through the front wall." Suddenly, Miles "became aware we were being watched." They looked up at a nearby snowbank and saw, poking out of the snow, "a bear's face blinking at us in the bright sun."

The face retreated into the snow, and Miles and Salisbury set up their cameras in readiness for the bear's reemergence; but not until midafternoon did the head again appear, and then only briefly. By then the sun had disappeared and the wind sliced into them; Miles recognized that—not altogether unexpectedly, given that he had spent several hours sitting still in a half-standing ice hide at 35 degrees below zero—he was shivering almost uncontrollably. The team retreated to their base camp, where Miles clambered into a snug sleeping bag with a warm drink until, after a couple of hours, warmth returned to all but his extremities.

After a sound night's sleep, Salisbury and the now-recovered Miles returned to the ice house, but the bear poked her head out of the snow just twice in ten hours of observation. Finally, on April 1, they saw what they had come so far, and endured so much, to see. At around eight thirty that morning, as their cameras rolled, the bear's head poked out of the den's entrance, followed by those of three cubs, one slightly smaller than the others and all three paler in color than the yellowish white of their mother. The cubs' first sight of the outside "was greeted with terrified squeaks, and they looked down the steep mountainside with trepidation," Miles and Salisbury wrote later. "The female fussed over them, then walked a few confident paces down, whilst two of the cubs followed gingerly, sliding backwards, with claws clutching the snow. The other stayed in the den entrance and cried so loudly that the female returned, and suckled all three in the sun."

After the effort they had gone through to make it this far, the filmmakers hoped that the bears would tarry a day or two before leaving the den behind, but that afternoon, "the female emerged purposefully, and started down towards the sea ice, followed protestingly by the little cubs." Although disappointed, Salisbury and Miles followed the mother and cubs at a discreet distance as they headed out onto the sea ice and away from shore, "in the realm of the seal, and within sight of her first meal for four months. We bade them a fond farewell and watched their departure into a white haze, feeling very sad it was all over."

Miles and Salisbury confessed to feeling quite emotional at being able to conclude their traumatic odyssey by witnessing, unnoticed by the bears they observed and far from any other humanity, the sight of a mother bear and her offspring taking their first tentative steps together. It is an emotion echoed by Thorsten Milse in *Little Polar Bears*:

> When the moment finally arrives, I'm always gripped by excitement: a den in the middle of the snow . . . a small black nose . . . you hold your breath . . . and then a white face appears. Cautiously and with some curiosity, a polar bear cub peeks out. Behind the cub, the mother's head gradually appears. For that one moment, I forget about everything else around me; as if in a trance, I press the shutter release of the camera. The endless wait has been worthwhile, the many hardships are forgotten—and I am suffused with a feeling of indescribable happiness.

Milse's book is a compilation of beautiful images of bear mothers and cubs in the Wapusk National Park in Manitoba, home to some 1,200 den sites, the principal denning area for the bears of Hudson Bay, and the largest denning area in the world. Although most dens worldwide—such as the one filmed by Salisbury and Miles—are within a few miles of the coast, for the bears that den in Wapusk, it is a forty-mile trek to the ice of Hudson Bay.

"Once she comes out of her den, she doesn't automatically leave right away," says Mike Spence. "The cubs need to become adapted to the environment. They've got a long journey ahead of them. It all depends on what condition the cubs are in. It depends on what condition she's in. They'll play, they'll get their muscles going, and then one day she'll decide it's time to go, and off she'll go."

Tom Smith has found that in Alaska, mothers and cubs tend to tarry at their dens on average two days before heading out for the sea ice, although some do so on the same day they emerge. During that time, if Smith and colleagues are able to watch the bears outside the den for one hour out of every twenty-four, they consider themselves fortunate. Often, he says, bears "will spend mere minutes per day outside the den. But then, why go outside? The environment is hostile and being outside also signals where the den is. They almost never nurse outside but again, why bother?"

Smith notes that staying at the den is both costly, in that it prevents the mother from hunting for food, and dangerous, as it advertises the den's location. At the same time, to leave too early, before the cubs are ready, would condemn them to certain death. "I'm convinced that the only reason mothers tarry at dens is to monitor cubs' growth and development," he says. "Once it meets some standard written in her genes, off they go."

The mother, desperately depleted after four months without food, nudges the young cubs onward, but for tiny legs that have hitherto been accustomed only to a small cave in the snow, it is a difficult journey. At times, particularly when danger may be near, the cubs clamber onto their mother's back. The travelers repeatedly stop and rest, the cubs nursing from their mother's milk and all three taking advantage of opportunities to nap and gather their strength. Eventually, they reach the shores of Hudson Bay, and the cubs gingerly follow their mother as she strides out, away from the beach and across the fractured and mangled ice, in search of seals to break her long fast.

Even as they reach the ice, danger abounds. Although the rate appears to vary by location, studies have shown that as few as 45 to 65 percent of polar bear cubs survive their first year. In Hudson

Bay, where an abundant spring is followed by the completely barren months of late summer and early fall, the survival rate of cubs may be as much as 20 percent below that of cubs in the Beaufort Sea.

The first few days are especially dangerous. It is a jarring conversion from the quiet warmth of their den to the subzero temperatures of an Arctic spring. Sea ice terrain is uneven and treacherous. And along the coasts and on the ice, danger lurks.

Adult male polar bears have been known to kill and eat young cubs, although how often they do so is unknown, and why they would do so is unclear. It seems unlikely it would be a purely predatory action: a newly born cub would not provide much energy, skewing the risk/reward ratio heavily toward the risk, particularly given the certainty of attack from an angry, protective mother. Perhaps, it has been speculated, there is a cold calculus involved: killing the cubs of another male decreases the competition and increases the survival chances of a male's own progeny. But that supposes that the male, having walked away from the female upon insemination, is able to distinguish between his cubs and those of others. Maybe, then, it is a means of improving his mating opportunities: as long as a female has dependent cubs, she does not enter estrus and become sexually available; without them, she is soon sexually receptive. But while such an explanation may make sense for brown or black bears, which have relatively small territories and could be expected to keep track of a female over the few days after she loses her cubs and before she enters estrus, it does not so readily for polar bears, whose ranges are considerably larger. Indeed, Steven Amstrup notes that on the two instances of infanticide he has observed in the Beaufort Sea region, male and female were already dozens of miles apart the following day and traveling in opposite directions.

Perhaps, speculates Amstrup, cannibalism of cubs doesn't actually serve any particular purpose at all. Perhaps it is a behavioral remnant from when polar bears were terrestrial, an atavistic anomaly that polar bears have not excised from their wiring. Ian Stirling, for one, thinks it happens only rarely and opportunistically. Yes, females do show a tendency to keep their cubs away from males, but they would be maternally remiss not to shelter them from hungry thousand-pound carnivores. And while it is not uncommon for males to move toward females with cubs, it is less common for them to be able to catch up to them, the cubs jostled along by their protective mother and all of them capable of maintaining a faster pace for longer than the heavyset males, which are large, well insulated, and quick to overheat when they run. When males do succeed in grabbing a cub, it is normally after surprising a family that is sleeping; if the mother is alert and able to respond, she will fight to protect them, even though the male is far larger. On occasion, it seems, she will do so even to the death: in two separate instances in 1984, researchers came across a male feasting on the carcass of a female, as her cubs cowered in the distance. In both cases, the scientists deduced that the male had killed the female; whether or not he went on to kill the cubs, they could not have survived for long.

Some cubs do not even make it as far as the sea ice, and ironically their mothers may be responsible for more instances of infanticide than any males. The physical strains on a pregnant female are immense, sometimes too immense. If, while preparing to make a den or even after settling into it, a female is not strong enough to see a pregnancy through to its conclusion, her body simply reabsorbs or aborts the fetuses, and she emerges from her den ready to resume sexual activity with the arrival of spring.

For some females, the physical limits are reached only when the cubs are born. By then it has been a long winter — for bears in Hudson Bay, eight months have already passed since their last meal — and the extra burden of rearing the newborns is a step too far. A 1985 paper in the journal *Arctic* cited seven reliably documented instances in which a mother was so malnourished that she killed and ate one or both of her cubs in order to survive. As Ian Stirling has written, "If there are seven documented instances of an event as difficult to observe as this, it may occur reasonably frequently."

The urge to survive, the overpowering drive to perpetuate her genes, that perhaps appears so cold when the polar bear mother devours her own young in order to increase her own chances of survival and thus of ultimately successful reproduction, seems to human eyes touching and warm when mother and cubs emerge from the den. She keeps them as close to her as she can: nudging them with her muzzle as they stand by her paws like uncertain children holding their mother's hands, or eager puppies keeping obediently to heel. It is a wonder that the mother does not trip over her offspring on a frequent basis, so tightly do they stick by her, winding around her paws, rubbing their heads against her legs. In time, they will grow up to be among the largest and most fearsome predators on Earth; now, they are vulnerable, largely defenseless, and insecure. Like any youngster taking its first steps in a frightening world, they want and need to stay close to their mother.

They will stick together, on average, for two years after emerging from the den, during which time their life will be, for all intents and purposes, a picture of familial bliss. The youngsters will tumble and play, staging mock battles, rolling over each other in the snow as their mother watches over them warily. They will curl up with her in pits in the snow, to sleep and suckle. And they will follow her closely, watching and tracking her every move, and for all that period of time, she will suckle them, even as she teaches them to hunt the seals that will constitute their diet for their entire adult lives.

During their first summer, the cubs will do almost no hunting of their own. They will, however, watch their mother closely and follow behind, sniffing where she sniffs, looking where she looks, lying down and waiting patiently when she begins to stalk or wait for prey. Occasionally, the waiting will prove too much, and the cubs, bored, will either pad up to their parent to see what is happening or will turn their attention to each other, biting, cuffing, rolling around, and running back and forth on the snow and ice—behaviors that sometimes earn a sharp rebuke from a mother anxious that her opportunity for a meal will be extinguished by such loud tomfoolery.

By the time the cubs are a year old, they have begun to hunt for themselves more often, but both the time spent doing so and the success rate of their efforts are substantially less than their mother's. With the passage of one more year, they have become significantly more adept: whereas yearlings and two-year-olds spend between 4 and 7 percent of their time hunting (as opposed to between 35 and 50 percent for their mothers), the former generally succeed in catching, on average, one seal every twenty-two days, while the latter have improved their success rate to one seal every five to six days.

Clearly, in that extra year the cubs have learned a great deal more from watching their mother, observations that have enabled them to become significantly more successful in their hunting. Their improvement is aided also by their growth, which gives them the extra strength needed to be able to kill and overpower a seal, and which is still fueled primarily by the nutrients from their mother's milk. That milk gives them strength, but its constant availability lessens the incentive for the cubs to refine their hunting skills with the rapidity that might be desired.

And so, once the cubs have reached roughly two and a half years old, once they have made it through their second winter in the open and grown big enough that they have enough fat reserves to survive the immediate future, they are forced to strike out by themselves.

Three years after she last mated, the cubs' mother is once more entering estrus. For the first time since conceiving, she is sexually receptive. The scent she exudes advertising her status attracts mature males from miles away, who now begin descending on the family's location. For the cubs it is a new and uncomfortable experience. For as long as they have been alive, their mother has protected them, kept them a safe distance from any inquisitive male. Now she welcomes her suitors' advances and rejects her cubs' attempts to seek sanctuary from their presence.

Scared, the cubs retreat from their mother and the approaching male. Confused, they look longingly after the retreating figure that has been the central focus of their entire lives. They start toward her, stop, look anxiously in the direction of the new arrival, and hesitatingly move away.

Finally, instinctively, they understand. They pace nervously back and forth, and they subconsciously move closer to each other as they moan uncertainly. The first phase of their lives, in which they relied almost entirely on their mother for nourishment and protection, is over. Now they must begin the second phase, the most dangerous few years of a polar bear's life. In the wind and snow, they huddle closer.

For the present, they have each other.

Otherwise, for the first time in their young lives, they are entirely alone.

Bear

The warming rays of the sun have long since dipped beneath the horizon; the cold and the dark hold the Arctic in their thrall, and the sea has frozen into a largely solid mass.

Across the Northern Hemisphere, the sun's gradual withdrawal heralds a gathering calm, a comparative stillness as a season of plenty yields to one of paucity. Summer's kaleidoscope of colors and cacophony of bird song surrender to winter's monochrome quiet. Priority is placed on the conservation of energy, rather than its exuberant expression. Resources are at such a premium that many mammals choose to enter a form of stasis, a metabolic depression known as hibernation, in which the body temperature drops and breathing slows. It is commonly assumed that among those species that hibernate are North America's bears, but the assumption is inaccurate. Both grizzlies and black bears do submit to periods of prolonged sleep during which they can, if undisturbed, sleep without stirring for much of the winter. But their body temperature falls only a few degrees, and they can be easily roused. It is not true hibernation.

The winters from which brown and black bears shelter are mild compared to the one that two young polar bears are about to experience, but for them there is no hibernation. In fact, winter, while lacking the abundance of spring, is not for polar bears the harshest season; in contrast to the experiences of their relatives, that distinction falls to summer, when ice is at its minimum, water is at its maximum, and the seals on which they prey are hard to reach. For these two bears, the onset of winter is a good omen; as they huddle against each other for warmth, looking not as fat as perhaps polar bears should, they could, were they able to consider such things, comfort themselves with the knowledge that the worst was over. They had survived their first summer alone. Now they needed only to make it through the long polar night, and soon they would be witnessing the new dawn of spring.

As they bed down, one of them lifts his head into the air. He is sniffing, thinking that he has perhaps caught a tempting smell drifting on the wind. As he sniffs, he does not see, in the clear night sky, the stars twinkling far above, or that one star appears to stand almost overhead hour after hour, night after night, seemingly never moving even as the others circle perpetually around it.

In recognition thereof, it is dubbed Polaris, the pole star; and in the imaginative menagerie of celestial creatures compiled by ancient, wondering eyes, it is at the tail of a small bear, which we know by the Latin name of Ursa Minor. Nearby, another loose assemblage of stars that in reality are billions of miles apart have been stitched together to create Ursa Major, the Great Bear;* and while the polar bears that stride across the ice below are the unquestioned rulers of the northern realm, it is Ursa Major after which their kingdom is named.

The ancient Greeks, analytical observers of worlds both celestial

---

* Today, this constellation is recognized primarily by its seven brightest stars, known in Britain as the Plough, and in North America as the Big Dipper.

and terrestrial, determined that Earth was spherical, not flat. They noticed that Earth tilted in relation to the sun, offering first one hemisphere and then the other over the course of the year, and deduced that, because of that tilt, there had to be a line beyond which the sun neither set at the height of summer nor rose in midwinter. They calculated correctly that this line was 66 degrees north of the equator (and that there was an equivalent line 66 degrees south), and when they projected this line onto the celestial sphere that they imagined surrounded the earthly one, they noticed that it grazed the constellation of the Great Bear. To the Greeks, that constellation, like the species after which it was modeled, was called *Arktos*, and so the region that lay beneath it was the land of the bear, *Arktikos*.

Furthermore, the reasoning continued, if there was a land to the north that lay beneath the Great Bear, then there must also be a counterbalancing landmass to the south. And if the northern lands were *Arktikos*, then those to the south must be the opposite — *Antarktikos*.

The ability to predict the existence of the Arctic and Antarctic did not necessarily translate into an ability to portray them with any precision. Writing roughly two and a half thousand years ago, for example, the poet Pindar imagined a people called the Hyperboreans who lived "beyond the north wind." "Illnesses cannot touch them," he wrote, "nor is death preordained for this exalted race." Several hundred years afterward, the Roman geographer Pomponius Mela more accurately proposed that both the northern and southern realms were frigid, that each was flanked by a belt of more temperate climes, and that north and south were divided by an impassable torrid zone.

(Centuries later, this latter view and its subsequent variations ran headlong into Christian orthodoxy. If the southern land were unreachable because of the torrid zone, then its inhabitants could not be the descendants of Adam and Eve. Besides, Christ had commanded the Apostles to "go into all the world, and preach the gospel unto every creature"; the Bible did not record that they had visited this land;

therefore they had not been there; therefore it couldn't exist; therefore the world must not be spherical; therefore it must be flat.)

The Arctic and Antarctic have many things in common, not least the fact that, as Pomponius Mela correctly surmised, they are both, relative to the rest of the surface of the Earth, very cold—although the Antarctic is more so than the Arctic. Both, as also predicted, have periods of uninterrupted daylight in summer and seemingly endless night in winter. Each region surrounds its respective geographic pole: the North Pole in the Arctic, the South in the Antarctic.

But there are also differences.

Antarctica is a frozen continent in the Southern Hemisphere, surrounded by ocean. The Arctic is the northernmost ocean, encircled by land. Both regions boast whales and seals in their waters, but Antarctica and its environs have penguins, and the Arctic does not.

And the Arctic, unlike the Antarctic, has polar bears.

Polar bears are found only in the north, penguins only in the south; cartoons and Christmas cards notwithstanding, the paths of the two do not cross. Should a population of polar bears be picked up by a pan-hemispheric tornado and deposited on Antarctica, their stay would assuredly be brief; they would likely rampage through the resident seal and penguin populations during the summer but find themselves bereft of sustenance during the long, hostile Antarctic winter—when seals retreat to the sea and the coasts of sub-Antarctic islands and temperatures plunge to levels that would challenge even this hardiest of predators.

Defining the Antarctic is somewhat easier than delineating the Arctic. The conventionally acknowledged northern limit of the former is the Antarctic Convergence, where the cold waters of the Southern Ocean clash with warmer seas farther north and establish a genuine boundary between temperate and polar realms. In contrast, the outermost reaches of the Arctic are more land than sea, and accordingly there is no uninterrupted circumpolar current to provide similarly convenient demarcation.

Distance from the equator alone is an insufficient criterion for inclusion. For example, the Arctic Circle (the line at 66 degrees north of which the ancient Greeks were aware) excludes Iceland and large parts of Siberia; Berlin, which is plainly not an Arctic environment, is on the same latitude as the Siberian town of Irkutsk, which suffers winter temperatures as low as −40°F and therefore arguably is. Many authorities use as a boundary the 10°C isotherm, a term that neither trips off the tongue nor is intuitively simple to grasp. It refers to the line along which, during the warmest month of the year (usually July), the average temperature is no higher than 10°C (approximately 50°F). That may seem somewhat arbitrary, but it does roughly correspond to the tree line, the point beyond which forest yields first to taiga (a region of scattered trees and scrub) and thence to tundra.

But there is another measure of greater relevance to our purposes: the southern limit of the winter pack ice. This encompasses all the waters that regularly freeze during winter, all the way south to James Bay in Canada, which is on the same latitude as England but is nonetheless, in terms of climate and ecology, Arctic in nature. It includes also Hudson Bay, Baffin Bay, Davis Strait, and Baffin, Banks, and Victoria islands in Canada; the coasts of Greenland; Svalbard and parts of the Barents Sea; Novaya Zemlya and the islands and coast of Siberia; the Kara, Laptev, and Chukchi seas off Russia; and the Bering Strait and Beaufort Sea off Alaska.

Not by coincidence, this is the range of the polar bear, *Ursus maritimus*, sea bear by scientific name and ice bear by nature. It is here, on the ice that covers bays and inlets, on the ice that is anchored to the coast, on the ice that drifts en masse offshore, on the ice at the Arctic's fringes, and even on the ice that encircles the North Pole, that the polar bear makes its home. It seems, to human eyes at least, desolate, a barren, hostile wasteland. Out of sight, however, below the feet of the prowling predator, the waters of the Arctic are rich in nutrients and life, some of which must occasionally come to the surface of water and ice to breathe and rest. It is in search of that life that the polar bear endlessly patrols its domain, a realm that it has made its own, to which it has, in a relatively short space of evolutionary time, become supremely well adapted.

To find a clue as to how that came to be, how, of all the bear species that live now or have existed in the past, just one should have ventured out onto the ice to hunt seals, we look to Alaska—not the icy coastal waters where polar bears roam today, but a very different environment indeed.

Of the many ways to explicate the scale of Alaska—the fact that it is twice the size of Texas, two and a half times as big as France, and almost six times as large as the United Kingdom; that it contains nineteen of the twenty tallest mountains in the United States, and the tallest in North America; the fact that its 35,000 miles of coastline, if somehow stretched into one long line, would encircle the equator with room to spare—the sheer diversity of its environments is one of the most impressive. The same state that in its northernmost reaches boasts tundra and sea ice is, in its southeastern region, lush and forested and regularly jammed with giant cruise ships.

There are no polar bears in southeastern Alaska, but there are brown, or grizzly, bears. There are many of them, in fact, particularly on three large islands—Admiralty, Baranof, and Chichagof, known colloquially as the ABC Islands—that form the northernmost extent of a 300-mile-long chain of mountainous island peaks called the Alexander Archipelago. There are three times as many brown bears as humans on Admiralty Island, 1,600 or so over an area of approximately one million acres, the highest density of brown bears in the world.

The grizzly bears of the ABC Islands are predominantly brown, with outer hairs that are often tipped with white or silver, giving them the "grizzled" appearance from which their nickname derives. They possess the large hump of muscle over their shoulders that distinguishes brown bears from other bear species. They eat roots and berries, fish, and small mammals. The males stake out territories and defend them fiercely from rivals. They look and act, in other words, like any other grizzlies.

But inside, they hold a secret.

The DNA of brown bears on Alexander Archipelago is different from that of brown bears anywhere else in the world. So distinct, in fact, is the DNA of the Alexander Archipelago bears from that of the rest of the world's grizzlies that other brown bears aren't even their closest relatives.

The Alexander Archipelago brown bears are most closely related to polar bears.

But we are getting ahead of ourselves.

We should begin at the beginning.

Polar bears are carnivores, and not only in that they eat meat. Indeed, they are the only truly carnivorous bear, as the other seven extant species are omnivorous or, in the case of the giant panda, effectively vegetarian. But they, and their bear brethren, are carnivores also in terms of their place in the great mammalian order of things.

All bear species alive today are members of the family Ursidae, in the order Carnivora, an order that includes the bulk of predatory mammals. (By way of comparison, human beings, along with bonobos, chimpanzees, gorillas, and orangutans, are generally—although not universally—grouped together in the family Hominidae within the order Primata, the latter of which also contains gibbons, monkeys, marmosets, and lemurs.)

The cat in your lap is a carnivore (in both the dietary and taxonomic senses), as are its larger relatives in the plains of Africa, the canyons of North America, and the jungles of South America and Asia. The family dog that sleeps at the foot of your bed, the raccoon that tips over the trash can at night, the skunk in the woods: all are members of the order Carnivora and thus relatives—second cousins, if you will—of the polar bear. Some, however, are more closely related than others. Generally speaking, cats and hyenas constitute one branch of the Carnivora family tree, and the remaining species—martens, minks, mongooses, weasels, wolverines, badgers, dogs, otters, seals, sea lions, and walruses—sit alongside bears on the other branch.

The progenitors of the Carnivora took their bow about 55 million years ago, around 65 million years after the appearance of the earliest true mammals and roughly 10 million years after a well-placed asteroid or two ushered the dinosaurs off the stage and allowed our mammalian ancestors—which, to that point, had mostly peered nervously at the proceedings around them from beneath bushes—to emerge blinking into the open.

Those progenitors, the miacids, were civetlike creatures, with long bodies and tails, that were likely mostly arboreal and probably fed primarily on insects and insect-eating animals like shrews. The Carnivora evolved from the miacids somewhat over 40 million years ago, and once they became established, they rapidly diversified. Within 5 to 7 million years, the dog, cat, weasel, and mongoose families had developed; and a little more than 10 million years after that—about 22 million years before the present—came the first bear.

Dubbed *Ursavus elemensis* and sometimes called the "dawn bear," it is known to us only from fragments of teeth and jaws. That is enough for paleontologists to determine it was about the size of a small terrier, but as to its appearance and habits we can only speculate. We can also only infer from an incomplete fossil record how it spawned other species and which species, in turn, evolved from those. We do know that by about 10 million years ago, *Ursavus* had disappeared, presumably as a consequence of climatic changes, the subtropical Europe in which it had evolved having become drier and the forests for which it was adapted having given way to steppes, plains, and desert. But where *Ursavus* fell by the wayside, others—evolving from either the dawn bear or contemporaries—emerged, spreading from Europe into Asia and thence the Americas. Some of the species that branched off from these lines have survived to the present day, while others, such as the cave bears of Europe and North America, and the giant, long-legged, short-faced bear—which at 2,200 pounds would have been approximately twice the size of most male polar bears—have long departed.

Today, eight species of bear—giant panda, sun, spectacled, sloth, Asiatic black, North American black, brown, and polar—are spread among four continents (Africa, Australia, and Antarctica being bearless). Each of them, in its own way, meets the more obvious visible criteria for a bear: a furry, essentially tailless, and somewhat sturdy body; wide paws with prominent claws; and a relatively large head with rounded ears and a muzzle less pointed than that of wolves. But within that broad generalization there are many differences, in size, shape, appearance, habitat, diet, and behavior.

Bears are generally caricatured as very large animals, and some of them truly are. But while the polar bear, the largest surviving bear species in the world, weighs in at up to 1,500 pounds or sometimes even more, the smallest, the sun bear—a denizen of the rainforests of Southeast Asia—is no bigger than a medium-sized dog and is in fact known in Thailand as a "dog bear" for that very reason. (Its more common name derives from a crescent-shaped patch on its chest.)

Bears' coloring ranges from the white appearance of the polar bear, through the black and white of the panda, to the darker pelage of the spectacled, sun, sloth, and appropriately named black bears. Their habitats extend from the Arctic Ocean surrounding the North Pole to the rainforests of Asia and South America and the river valleys, mountain forests, and meadows of North America and Europe.

Their feeding habits are similarly varied. The sun bear spends evenings scurrying along the rainforest floor in search of food and its days nesting in the branches of trees; across the Pacific, much the same is true of the Andean spectacled bear, the only extant bear species in South America and, in the view of many scientists, the closest living relative of the extinct giant short-faced bear. Feasting on anything from fruits to insects to small animals, both species are opportunistic feeders, as are brown and black bears, which will eat fish and carrion but also consume plants and berries. Polar bears, in contrast, are truly carnivorous, their diet dominated by seals; the giant panda is famously all but reliant on thirty species of bamboo; and the shaggy-haired sloth bear of the Indian subcontinent, Nepal, and Sri Lanka is almost exclusively insectivorous, consuming in particular vast amounts of ants and termites, using an extraordinarily powerful force of suction to remove them from their hills and mounds.

The sloth bear acquired its name after the first skins reached Europe in the early eighteenth century. Its long claws and snout and shaggy mane led to the assumption that it was, in fact, a bearlike sloth, and it was initially so categorized. Even when the mistake was corrected and the species was properly classified as a bear, its common name stuck. It is not the only such example of bear-related confusion and uncertainty.

Although popularly referred to as bears, koalas, for example, are in fact marsupials that happen to boast certain physical traits commonly considered to be bearlike. Nor is the red panda, which shares with its larger namesake distinctive facial markings and a fondness for bamboo, a bear; rather, it appears to be more closely related to the raccoon.

For some time, there was considerable debate as to whether the great, or giant, panda was itself truly a bear. Its size suggests that it is—and indeed it was immediately identified as such by Père Armand David, the French priest and naturalist who was the first European to spy a panda pelt, and subsequently an actual panda, in 1869. But a panda's head and jaw are shaped more like a raccoon's, and whereas most bears are omnivorous with a tendency toward the carnivorous or insectivorous, pandas are almost exclusively vegetarian. Taxonomists debated for years whether the giant and red pandas were closely related, whether they should be grouped with raccoons or merited their own family. Ultimately, genetic studies demonstrated what anatomy alone could not: giant pandas are indeed bears—are, in fact, the oldest of the bear species currently on Earth, having split off from the main ursid line about 10 million years ago.

Because of the giant panda's unique qualities and early divergence, taxonomists classify it in a subfamily, Ailuropodinae, of which it is the sole member. The spectacled bear also occupies its own subfamily, the Tremarctinae. The six remaining species have a more closely shared ancestry, having all diverged from the same branch of the family tree.

Before *Ursavus* departed this Earth for good, it begat *Protursus simpsoni*, which in turn led to *Ursus minimus*. The first true ursid, *Ursus minimus* was the approximate size of today's sun bear, although over the course of a few million years it apparently grew in size before giving rise to the yet larger *Ursus etruscus*, which was roughly equal in stature to the American black bear. *Etruscus* would in turn lead directly to the now-extinct cave bear of Europe, but it would also prove to be the granddaddy of six of the extant eight bear species: one branch led to the sun bear, one to the sloth bear, another to the Asiatic and then American black bears.* Then, between a million and a million and a

---

* Or something like that. A 2004 analysis of bear phylogeny concluded that precisely where sloth and sun bears fit into the scheme of things, and when they diverged from the ancestral branch, remains uncertain.

half years ago, the brown bear made its maiden appearance, first in Europe, then Asia, and ultimately North America.

Finally, probably around 200,000 years ago, at much the same time as *Homo sapiens* was emerging in Africa, a new bear species took the stage. Clues to when, where, and how that came about are found in the brown bears of the Alexander Archipelago.

Studies have shown that the mitochondrial DNA of bears tends to change by about 6 percent every million years or so. The contention, repeated above, that brown bears split off from black bears around a million and a half years ago is supported by the observation that their mitochondrial DNA differs from that of black bears by between 7 and 9 percent. Polar bear DNA diverges from that of most brown bears by around 2.6 percent, which on its own might lead to the conclusion that polar bears evolved from brown bears a little under a half-million years ago. But the divergence between polar bear DNA and that of the Alexander Archipelago grizzlies (which themselves, according to the genetic clues, separated from the brown bear lineage between 550,000 and 750,000 years in the past) is a mere 1 percent, suggesting that their appearance is much more recent.

Any scenario to explain the sequence of events leading to the divergence of the Alexander Archipelago bears and the emergence of polar bears is inevitably and necessarily speculative. But thanks to the genetic evidence, such speculation is at least informed and allows us to paint a picture that looks something like this:

Half a million or so years ago, a stock of coastal brown bears in or near the Arctic became isolated, perhaps by surging glaciers or advancing sea ice. Sometime during the ensuing millennia, they began venturing out onto that ice, probably initially preying on young seals that had never had cause to fear predators, and perhaps then graduating to adult ringed seals at their breathing holes. Over time, natural selection presumably favored those bears with lighter coats for camouflage, with teeth better adapted for tearing at meat, and with other physical adaptations such as larger feet for easier swimming between ice floes.

It all happened in the blink of an evolutionary eye. When the last Ice Age began, there were no polar bears. By the time the most recent glacial period ended, polar bears as we know them had become established. The other descendants of the original, isolated brown bear stock that gave rise to them presumably either were reabsorbed into the broader grizzly gene pool or died out. Except, that is, for those on the Alexander Archipelago.

When ice advanced southward into what are now more temperate latitudes, the islands of the archipelago pierced the glacial shroud, acting as what biologists call a refugium, a safe haven where wildlife populations were able to endure. Those populations were isolated by the surrounding ice and then, when the ice retreated, by the passages of water between the islands and the mainland. Which is how it came to be that the archipelago boasts, for example, its own subspecies of dusky shrew and northern flying squirrel, even an Alexander Archipelago wolf — and, of course, the ABC Islands' brown bears, which today contain, deep within their cells, echoes of a distant past and the birth of a new species.

There is not an abundance of polar bear fossils — animals that live on sea ice tend to die on sea ice, and animals that die on sea ice tend to sink to the bottom of the ocean when the ice melts — but those that do exist suggest that early polar bears were larger even than the ones that wander the Arctic today. They were, in the words of one researcher, "gigantic."*

---

* Indeed, mammals generally were larger during the Ice Age. This was the time of, among other species, the giant beaver, the giant sloth, and the Irish elk, the largest deer that ever lived. That said, some doubt has more recently been cast on the accuracy of the assumption of early polar bears' gigantism.

(Given the greater size of those early bears, a 1971 study that concluded average skull size was larger in polar bears in the Chukchi Sea between Siberia and Alaska was cited as evidence that this must have been where the species originated. But although the origination location may be correct, the notion that there is a gradient in skull size almost certainly is not. The study was based on skulls housed in museums around the world, but the sources of those skulls were not uniform; it is believed that many of those from the Chukchi Sea, for example, may have been donated by trophy hunters, who were of course adept at, and focused on, hunting larger bears.)

But if modern polar bears are smaller than their predecessors, they are still the largest bear species in the world. The largest male weighed in Hudson Bay was a thirteen-year-old that tipped the scales at almost 1,450 pounds; the average weight for mature male bears in that population is approximately 1,100 pounds. Steven Amstrup observes that the heaviest bear he and his colleagues have weighed in the Beaufort Sea area of Alaska was roughly 1,350 pounds but that some tranquilized animals were so enormous that they could not be lifted up with a weighing tripod, or even by helicopter. Doug DeMaster and Ian Stirling have estimated that some weigh as much as 1,760 pounds. Females are significantly smaller, and their maximum weights rarely exceed 880 pounds even when they are at their most obese, when they have gorged following mating and prior to denning.

A mature male may measure over nine feet from the tip of his nose to his stubby tail, stand five and a half feet at the shoulder when on all fours, and reach thirteen feet into the air when standing on his hind legs. Polar bears are not only the largest bears in the world today, they are very nearly the largest that have ever existed; only their ancient ancestors and the extinct giant short-faced bear were larger.

Size, however, isn't everything. There is more to being a polar bear than being big.

•   •   •

It all begins with the head.

If you had never before seen a polar bear, never even knew that such a thing existed, but were nonetheless familiar with other things ursid, from a panda bear to a teddy bear, Yogi Bear to Gentle Ben, the moment you saw the face of a polar bear you would almost certainly know instinctively and immediately what type of animal it was. It is unmistakably . . . *bearlike*, with its rounded ears and furry muzzle. But compare the face of a polar bear to that of a grizzly and one difference is immediately apparent: the latter's seems flatter and wider, while the former's tapers into more of what might be called a Roman nose. The largest brown bear skulls are larger than the largest polar bear skulls; although they are in fact only marginally wider in relation to length than those of polar bears, they are also higher, creating more of a "dish-faced" impression than the more elongated visage of a polar bear. The more streamlined silhouette is accentuated by the neck, which in a polar bear is relatively long and slender, an adaptation that, in combination with the narrower skull, enables polar bears to succeed in hunting prey that would be beyond the reach of their evolutionary cousins.

"If a grizzly sticks its head in a seal hole," points out the San Diego Zoo's JoAnne Simerson, who has spent many years studying polar bears in captivity and the wild, "then it isn't going to be able to pull it back out."

And when a polar bear seizes a seal, the weapons it contains inside that elongated skull allow it to dispatch and dissect it with consummate efficiency.

Its forty-two teeth are significantly different in size, shape, and composition from those in a brown bear's jaw, and they are weighted far more heavily toward grabbing and holding prey and shearing meat. For example, the first premolars, the teeth immediately behind the pointed canines, are vestigial, effectively creating a gap between the canines and the molars that allows the former to penetrate deeply into seals and other prey without hindrance from adjacent cheek teeth.

The ears, too, are different from those of a brown bear. A grizzly's ears may not be its most prominent feature, but they are significantly larger than those of its nearest relative (while those of a sloth bear appear positively elephantine in comparison). The smaller the ears' surface area, the lower the amount of heat lost through them; and while losing heat is an occupational hazard of Arctic life, it is a danger to which a polar bear is tremendously well equipped to respond.

Any energy-conscious homeowner will agree that a sign of good attic insulation is a roof on which snow does not melt in winter. So it is with polar bears, the still-frozen snowflakes on their fur evidence that even as they maintain their core body temperature, which is almost identical to that of humans, the tips of their hairs may be as much as 75 degrees cooler.

By way of illustration of the effectiveness of polar bear insulation, Ian Stirling tells a tale of a fellow researcher who wondered if infrared photography could be used to detect polar bears on the ice, given that all warm bodies emit infrared radiation.

"To test the idea, he found a bear and took some pictures," Stirling wrote of his colleague. "The bear was so well insulated it gave off no detectable heat at all. But there was a spot on the infrared photo, just ahead of the bear's head . . . made by its breath!"

So effective, in fact, is a polar bear at thermoregulation that even in temperatures that would threaten hypothermia in humans, its bigger concern is staying cool. This is the primary reason why its pace seems so unhurried: at temperatures between −4°F and −12°F, a bear's body temperature remains fairly constant when walking at about two and a half miles per hour; by the time it is moving at a mere four miles per hour, its body temperature may soar to 100°F. No wonder, then, that although adult bears are capable of bursts of high speed when launching surprise attacks on basking seals, those bursts are brief and, particularly if the seal escapes, followed by exhausted recuperation.

When the ambient temperature is too high for comfort, polar bears respond by doing as little as possible, seeking out shade or shelter and lying still, expending no unnecessary energy and waiting for conditions to cool. Even the arrival of lower temperatures and the season's first ice and snow may offer only marginal respite.

"Typically what you'll see them do is spread themselves on the ice, so their groin area can cool off," says JoAnne Simerson. "What I've seen them do, before there's snow and ice but the ponds are beginning to freeze, is actually punch through the thin layer of ice on the ponds and lie down and sit in the water."

And, when sitting in water doesn't help, they can immerse themselves in it.

Polar bears are excellent swimmers. They can—and frequently do—swim for hours at a time, although too prolonged an exertion can result in severe exhaustion and the need to recuperate for lengthy periods, and may be particularly dangerous for young cubs. The same long neck that enables a polar bear to plunge its head through ice holes and under water when hunting also allows it to keep its head above water for long stretches when swimming; and the fact that it is able to swim for long stretches owes a great deal to what are by far the largest feet in the bear world.

At up to twelve inches in diameter—almost twice as big as a brown bear's feet—a polar bear's front paws are striking in their immensity; in the photograph at which I am currently looking, which I took on the shores of Hudson Bay a few weeks before writing these words, a resting bear is looking at the camera and at me, its muzzle lying on a paw that seems almost as large as its head.

"They're like snowshoes," says Simerson; "it's all about weight distribution," a means of spreading the impact of a bear's bulk to make it easier for it to walk over ice. Indeed, when a bear is setting out uncertainly over ice that is particularly thin, it will often all but spread-eagle itself to distribute its weight as evenly as possible. But such large paws are also valuable for fatally swatting seals. And they make magnificent oars with which a bear can doggy-paddle through the water, while its rear legs curl up underneath its body and act like rudders.

A polar bear's buoyancy is considerably aided by a thick layer of blubber, which not only helps keep a bear warm, but also enables it to store huge reserves of fat for lean periods and metabolize them when needed, a response to an environment of uncertain and uneven productivity. Polar bears frequently eat seemingly to excess when food is available and then slow down their metabolism substantially while drawing on fat reserves, a state known as "walking hibernation," when pickings are thin and food is hard to find.

That layer of blubber can be as much as four and a half inches thick and is topped by skin that is, perhaps surprisingly, black. Equally surprisingly, a polar bear's fur is not actually white.

It is in fact unpigmented, but because it reflects visible light it appears white to the human eye when it is clean and in sunlight. During sunrise or sunset, the fur may actually appear to take on the orange-yellow hues of the rising or setting sun; later in the season, in spring or late winter, before the annual molt that begins in April or May, bears may appear to be "off-white" or even yellowish.

While sunlight is reflected off the pelt (contributing to its white appearance), the late Norwegian scientist Nils Øritsland found that some of it passes through and reaches the black skin, which naturally enough warms up.

"Warm surfaces, of course, emit long-wave infrared radiation, or heat," notes Øritsland's colleague David Lavigne. (This is the principle behind night-vision optics.) "So the skin emits long-wave infrared and it turns out that, when it hits the hairs on the way out, they trap the heat inside the pelt."

In other words, polar bear hairs let some of the sun's warming rays pass through, but they don't let the heat radiate back out. Not only that, but because the hairs are hollow, they contain air that is warmed by the heat trapped within the pelt, causing further warming.

It is the final line of defense against the polar bear's most constant and potentially dangerous enemy, the bitter chill of its Arctic domain, and perhaps the most effective. (Although not, of course, in the very dead of winter, when there is no daylight at all. For the mechanism to work, it needs sunlight.) It is also a line of defense that has been frequently misreported and misunderstood.

A polar bear hair does not, as Richard C. Davids writes, act as "a miniature light pipe that funnels only ultraviolet light down through its core to be absorbed by the black skin." That idea — that polar bear hairs not only trap heat from sunlight but actively channel the ultraviolet directly to the skin — is a myth, but a persistent one. Its origins lie in a misreading of both Øritsland's research on polar bear fur and another study he conducted with Lavigne.

In 1974, the two men were attempting to determine a means of counting seal pups from the air that was more effective than searching for white seals on white ice.

"We discovered that ultraviolet photography turns white seals black, and in the course of doing so, we discovered that ultraviolet photography turns most white animals black," Lavigne recalls. "We were in a hurry and wanted to photograph a white animal from the air, so we went to Churchill, Manitoba, and photographed a polar bear with three cubs, and they all showed up black. That led to a cover story in *Nature* called 'Black Polar Bears.'"

The reason seal pups and polar bears look black when photographed in ultraviolet is that their fur absorbs UV radiation; conversely, they appear white to us in normal conditions because the fur reflects visible light. There is no obvious benefit to the bears — or the seals — from absorbing UV radiation in this way; as Lavigne noted, it turns out that most white animals do so, and later he discovered that the Canadian military's equipment, painted white to provide camouflage in snowy Arctic environments, did as well.*

But subsequently, some American military researchers who had not been involved in either study conflated the two findings and, speaking to a reporter for *Time* magazine, described the process differently.

"Now the ultraviolet hits the polar bear, gets to the hollow pipe which acts as a fiber optic, transmits this high-energy ultraviolet radiation into the skin, and then heats up the bear," says Lavigne, describing the distorted version. "But the problem is that the ultraviolet never gets to travel down the hair because the whole point of what Nils and I described is that it is immediately *absorbed* by the hair."

Even so, what Lavigne describes as "the myth of the solar polar bear" endures. "Solar energy? Do polar bears hold the secret?" asked the *Washington Post* rhetorically in 1987. "Polar bears' fur holds clue to better lasers," insisted the London *Sunday Times* in 1995. Even the august *Scientific American* got in on the act, headlining a 1988 article simply and succinctly, "Solar Polar Bears."

It endures partly because distinctions among visible spectrum light, infrared, and ultraviolet, among absorption and reflection and transmission, can be subtle and confusing. It endures because it is hard to fact-check when so many sources have repeated the story as gospel. But it endures also because it seems like something that should be true, that a magnificent animal in an almost impossibly hostile environment should have developed an almost fantastic mechanism for keeping warm.

The pity of it is that the truth needs no elaboration. Every polar bear is so well insulated that not only does it boast a thick layer of blubber and a warm pelt, but that pelt warms the cold air that reaches its skin and traps it against its body, enveloping the bear in a thermal

---

* As Lavigne points out, all kerotinized protein—that is, hair—provides an element of UV absorption. "If nothing else, it prevents sunburn," he notes.

blanket. Polar bears, in effect, are mobile, furry greenhouses. That, by itself, is remarkable enough.

As Robert Bieder has observed, the polar bear is a creature of paradox. It is a white bear with translucent hair and black skin. It is an enormous predator that walks softly and almost silently across the ice. It is an Arctic resident whose major problem is not staying warm but keeping cool. And the winter, when other bears are generally hibernating, is for the polar bear among the most active times of the year, as it stalks the ice in search of seals.

To the members of a sixteenth-century voyage of exploration, it was a white bear "of a monstrous bigness." Eighteenth-century whalers dubbed it the "farmer," tending the ice fields across which it wandered. Scientists call it *Ursus maritimus*, the sea bear. But it is in Norwegian or German—where it is known as, respectively, *ijsbjørn* and *eisbär*—that the nomenclature most accurately describes the species.

For these land carnivores that are officially classified as marine mammals are, above all else, creatures of the ice.

Ice

We were moving.

It was morning, I was in my bunk, I had been asleep, and when I woke up I could feel the movement of the ship. I lay there, blinked a few times, rubbed my eyes, stretched, gathered my thoughts.

We weren't supposed to be going anywhere. The previous day we had arrived off Barrow, the northernmost town in the United States, on the Beaufort Sea coast of Alaska. We had dropped anchor that afternoon; we were not scheduled to weigh it for several days, time for new crew to arrive and existing crew to leave, and to change the onboard complement of scientists and journalists.

But we were unmistakably on the move. And yet, something about our motion didn't feel right. It was almost as if we were moving . . . backward.

Frank Kamp, the chief mate, had been the first to spot the impending collision. He had been on anchor watch, alone in the wheelhouse early

in the morning, plotting our upcoming course on the chart, keeping
the bridge tidy, casting an occasional eye out the window.

The watch had been uneventful, as anchor watches often are. The
Beaufort Sea current was strong, driving at several knots along the
Alaska coast; but although there was some ice, it was not especially
plentiful. Floes appeared sporadically, mostly passing harmlessly by;
those that appeared to be on a collision course looked to be of little consequence, but even so, when necessary, Frank nudged the rudder just enough to lessen or avoid the blow. Not that the strengthened
hull could not withstand the impact, but there was no point subjecting
it to any more encounters with ice than absolutely necessary. Besides, a series of jarring, thumping impacts throughout the night and
early morning would have done little to endear the first mate to his
fellow mariners.

As early morning dawned — the transition from night to day barely
perceptible at this latitude in summer — the fog descended. Peering
through the binoculars, Frank scanned what passed for a horizon,
then paused. He lowered the binoculars, raised them again, wondering if his mind was playing tricks on him and finally realizing that he
truly was seeing what he thought he was seeing. There was indeed an
enormous ice floe headed for the ship, one that no amount of rudder
turning could avoid. This one was going to hit, and when it did it was
going to do more than briefly rouse a few people from their sleep.

He rushed below to fetch the captain, but barely had the pair made
it back to the wheelhouse than the floe struck. The force of a half-
million tons, one billion pounds of frozen water, drove into the *Arctic
Sunrise*, and after token resistance the ship acquiesced to its demands.
Now we too were headed along the coast at several knots, powerless
to deny the urgings of the enormous piece of ice that had made us
its plaything.

The severity of the situation was not initially apparent to all on board. For many it was an opportunity to photograph sea ice up close, the dirty, gnarled mass stretching in every direction and providing grist for photographic mills. There was even the added bonus of a brief appearance of the anchor chain—which, as captain and mate deployed the ship's thrusters in a desperate battle to twist and turn the *Sunrise* free of its imprisonment, abruptly sped across the surface of the floe, anchor in tow, before disappearing anew beneath the surface. The cook, awakened by the brouhaha, stuck his head out of his cabin porthole, only to be rapidly reprimanded by those assembled on deck, fearful that the anchor and chain might make an unscheduled reappearance and ricochet into the unwary.

But if not all those present immediately grasped the gravity of the predicament, Captain Arne Sørensen was all too aware of the dangers, not only because his years of sailing in polar seas had made him acutely familiar with the risks posed by marauding ice floes, but also because he had firsthand experience of their power to entrap the unsuspecting in their suffocating embrace.

On October 27, 1985, Sørensen, then captain of the Australian research ship *Nella Dan*, was grinding northward through the sea ice surrounding Antarctica, threading his ship gently through a maze of open water that promised to provide safe passage to the relative sanctuary of the Southern Ocean. As he did so, an easterly wind arose as if from nowhere, pushing the floes into each other on all sides, compressing them around the hull and under the bow. Almost in a matter of moments, the ship lifted clear of the water and ground to a halt. Over the next few days, the broken ice froze into one enormous field, and the *Nella Dan* was completely beset. There it stayed, frozen in place, for seven weeks, until released on December 14 by the Japanese icebreaker *Shirase*. The severity of the *Nella Dan*'s captivity was underlined by the fact that, having plowed without resistance through a seemingly endless landscape of ice on its way to perform the res-

cue, the *Shirase* was forced, upon arrival at the frozen Alcatraz that en-
cased the hull of Sørensen's vessel, to back up and charge through ice
that in one spot close to the Australian ship was measured at sixty feet
thick.

No fate so dramatic befell the *Arctic Sunrise* off Barrow that July day.
After two hours of frantic maneuvering, Sørensen extricated us from
our encounter, albeit not before we had been pushed four miles along
the Alaska coast. ("Farthest I've ever traveled on an anchor watch,"
quipped one of the mates.) We were fortunate. Seafarers' encounters
with sea ice have all too often ended far less happily.

Even Sørensen's enforced sabbatical on board the *Nella Dan* was relatively mild compared to the punishment the polar sea ice has at times
meted out to those who have dared to enter its domain. Indeed, a little over a month after the Australian vessel escaped its imprisonment,
another ship, the *Southern Quest*, was trapped between ice floes in the
Ross Sea; its hull ruptured, it sank within a matter of hours, forcing
the crew onto the ice, from where it was rescued by military personnel
from a nearby U.S. Antarctic base.

It need not always be thus. George Best, a member of a sixteenth-
century voyage in search of the Northwest Passage (a path from Atlantic to Pacific across the top of North America) and one of the first
Europeans to encounter sea ice, recorded a scene that was far more
placid than that experienced by the *Nella Dan* or *Southern Quest*, or even
by us that morning on the *Arctic Sunrise*:

> . . . many of our company lept out of the Shippe upon Ilands of Ise,
> and running there uppe and downe, did shoote at buttes upon the Ise,
> and with their Calivers did kill great Ceales, which use to lye and sleepe
> upon the Ise, and this Ise melting above at the toppe by reflection of the
> Sunne, came downe in sundrye streames, whyche uniting together, made
> a prettie brooke able to drive a Mill.

Best's positive portrayal of the surroundings was doubtless influenced not only by the novelty of his experience and the incredulity he felt at witnessing such a scene, but also by the sunny conditions and the fact that the expedition departed the Arctic before the onset of winter. It is a sense of wonderment shared by innumerable mariners in the years since; on a summer's day, with the sun's rays reflecting brightly off scattered, flat chunks of sea ice, the polar sea is a sponge that magically soaks up time and slows it to a crawl, a place of blissful calm and quiet.

But as the seasons pass and the sky darkens, the ice floes thicken and multiply, the amount of open water steadily diminishes, and the tenor of the surroundings becomes altogether more threatening.

By the time the sea surface is predominantly frozen, the summer silence has given way to the thunderous pounding of ice floes grinding together, a sound that Frederick Cook, who claimed to have been the first to the North Pole in 1909, described as "terrifyingly like a distant thunder of guns."

When a snowstorm drove ice against the hull of a vessel on a 1569 Dutch voyage of Arctic exploration, Gerrit de Veer, the ship's second mate, wrote that it was "most fearfull both to see and heare, and made all the haire of our heads to rise upright with feare."

De Veer's ship was ultimately destroyed by the ice, but it was neither the first nor the last to meet its end that way. More than thirty vessels were crushed during the long search for the Northwest Passage before the completion of its first traversal in 1906, and those that were not wrecked were frequently trapped. In 1829, the *Victory*, under the command of John Ross, was seized by ice west of Baffin Island and was not released for almost four years.

"Driven by the power of winds and currents sweeping across the pole, ice as hard as concrete and as high as a house is broken, tilted, piled, ground and refrozen into an impassable chaos, booming and groaning as if it were alive and suffering," writes Robert McGhee in his book *The Last Imaginary Place*. "The wooden ships of the age of exploration had no chance against such a force."

"It's like a wonderful time-lapse version of plate tectonics," explains Brendan Kelly of the University of Alaska, a biologist who was among those we were about to welcome on board the *Arctic Sunrise* when we were so rudely interrupted by the ice floe off Barrow. "You watch these ice floes come together, these massive plates, and one will subduct, one will layer over another, mountains result, and it all happens in frighteningly real time. It's pretty dramatic to see. I've watched ridges that are five, six meters high forming in front of my eyes. It's pretty intense."

Polar explorers Rune Gjeldnes and Torry Larsen, who in 2001 completed the first and so far only unsupported crossing of the Arctic Ocean, referred to the sea ice across which they labored as the "devil's dance floor." Eric Larsen, who with Lonnie Dupre twice (in 2005 and 2006) attempted to become the first to cross the Arctic Ocean in summer, and on the second attempt succeeded in reaching the North Pole, recalls, "For me, what I realized was that conditions were always getting worse. Just when I thought I had a handle on how the ice was behaving, there would always be some new way in which it was mashed, crumpled, rafted, pressured. Trying to cross it was crazy, just full-on crazy."

In such an environment, it is hard to imagine that anything could possibly live.

"There are pieces of ice that are heaved upward, there are pieces of ice that are ground up into smaller chunks and then heaved up again. There were times when it might take us two hours to go a quarter of a mile. It felt more like mountaineering at times. Then there are sections of open water that are newly frozen, so the ice is so thin that, for example, it was bending underneath our skis," Larsen says. "It's an environment to travel in that is only stress for any organism that is traveling across it by its own power, because there is never a moment to relax. It's always changing."

And yet, incredibly, in the midst of this grinding and churning, amid the desolate icy plains and the imposing mountain ranges composed of car-size chunks of ice, suddenly, out of nowhere, there will emerge not only life, but life at its largest and most majestic, a silent yet powerful predator supremely adapted to surroundings that would rapidly visit death upon any unprepared human.

"We had a bear come into our camp one day," Larsen recalls. "We scared it away by firing flares, and I followed its tracks for a ways. I could see how it had come up to a lead that was newly frozen, hopped on it, walked on the ice a little bit, fallen through, swum across, and I could see where it got out on the other side. For Lonnie and me to have done that, it would have meant stop, take our skis off, catamaran our boats together, put them in, chop ice, claw our way through that, get to the open water, and paddle a little way through that, then one of us would have got out, held the boat so the other person could get out, hauled the boats out . . . it would have taken us fifteen minutes.

"I would say that bear was able to make that trip in less than a minute."

Approximately 10 percent of the land's surface, and 7 percent of the ocean's, is at any one time covered with ice. The former consists of so-called alpine glaciers that squeeze through mountain valleys, inching forward at a speed that originated and defines the term *glacial*; and of ice sheets that at times have covered large swaths of the globe but are now confined to Greenland and, especially, Antarctica (the latter containing 90 percent of all fresh water on Earth). Ice sheets and tidewater glaciers calve icebergs, great chunks of ice that have built up year by year, century by century, often containing frozen water hundreds or even thousands of years old.

Icebergs, particularly Antarctic icebergs, can be truly massive. I have seen bergs that have stretched more than a mile in length, but such apparent giants are but mosquitoes compared to the largest ever recorded, which was more than seventy-five miles long and boasted a total surface area approximately equivalent to that of the country of Luxembourg.

Icebergs may be twisted or curved, upright or lengthy, small chunks the size of pianos, or grand islands of ice. They may have holes carved into them or even arches worn right through them as a result of months or years of impact from crashing waves. They may be white; they may contain straight stripes of brown, black, or green from sediment or algae that became entrapped during the bergs' formation; they may have patches of remarkable iridescent blue.

For mariners in polar seas, they may be welcome visitors in what can at times be a monotonous vista.

They are often extraordinarily beautiful.

They are also, in the context of this discussion, essentially irrelevant.

Caricatures and imaginings to the contrary notwithstanding, icebergs are not polar bear habitat. While it is not beyond the realm of possibility that a bear might clamber onto an especially low-lying, highly weathered berg, the energy required for an 800-pound animal, weighed down with sodden fur, to haul itself out of the water and up several feet of smoothly weathered ice would be considerable, even assuming it were able to secure an adequate grip. A weathered iceberg may be a last-gasp life raft for an exhausted swimming bear, or it could conceivably be a vantage point for a bear walking across the sea ice when a berg has become grounded and surrounded by closely packed floes. But in the latter scenario, icebergs are not truly members of the polar bear's environment but interlopers into it.

The ice that forms the realm of the polar bear is altogether more ephemeral than that found in glaciers and icebergs. Although some sea ice, especially in the Arctic, persists for several years, it is never, unlike glacier ice, centuries or millennia old. It is not created from the steady accumulation of precipitation over hundreds and thousands of years. It is not, in fact, fresh water at all. It is the frozen surface of the sea, a surface that freezes and melts in an annual cycle, its extent waxing and waning so that, in time-lapse satellite images, it looks almost like a lung expanding and contracting.

And like a lung, sea ice—its formation and melting—breathes life into the Arctic.

As the ice forms, it traps a multitude of microscopic creatures in the water, including unicellular algae called diatoms. More than 200 species of diatoms have been found living in Arctic sea ice and have become specially adapted to this most unlikely environment. Diatoms are plants; they must photosynthesize to survive, and they reproduce by dividing into two. And yet they can move, and move they do, taking advantage of channels in the ice to crawl toward the bottom of a floe. That may seem counterintuitive, given that in so doing they move farther away from the surface of the ice and thus any hope of receiving the sun's nurturing rays, but there is method to their movement. There is anyway little or no sunlight during the winter months; it is for spring that the diatoms wait.

As ice freezes, it expels much of the salt that was in the water from which it forms, so that a floe is less saline—"fresher"—than the sea in which it floats. When the sun returns, its rays penetrate the ice and warm the water beneath, melting the ice at the bottom of the floes—which, being fresher, is lighter than salt water and thus floats at the surface, diatoms included. Bathed in sunlight, the diatoms bloom, setting in motion an avalanche of feeding and breeding that makes spring in the Arctic an eruption of life of such force that it sustains much of the region's marine ecosystem for the rest of the year. (Arctic marine mammals have layers of blubber not just for warmth and buoyancy but for storage, so that they can stock up during the season of plenty and live off their reserves as necessary when harder times return and endure.) The focus of this explosion in productivity is the ice edge, where diatom blooms attract the attention of copepods and other phytoplankton, which are eaten by small fish, which in turn are eaten by arctic cod, which in turn are eaten by seals.

And the seals, finally, are eaten by polar bears.

Patience.

That is what their mother had taught them, what they had picked up on those occasions when they had watched her intently, before their attention had wavered and they had begun to roll around in the snow, a disturbance that was evidently enough to disturb the seal their mother had been intently eyeing and that resulted in a sharp reprimand, a rebuking hiss that once more sharpened their focus.

It had been difficult then to fully grasp the urgency of their education. Their mother's disapproval was discomfort enough to prompt greater dedication to the task for as long as their concentration held, but there was always the reassurance that, even if they failed in the task she had set them or even if occasionally their youthful shenanigans had interfered with her own efforts, her experience could almost always be relied upon to ensure a more successful conclusion the next time. And even if they did not dine daily on seal blubber, they were guaranteed the warm, calorie-rich nutrition of her milk. Every day for almost two years after their emergence from the den, even as they grew and became progressively more successful at killing seals for themselves, she would nourish them, leaning back against a snowdrift, cradling them to her as they nursed.

Now that security had been ripped away from them, its absence made all the more acute by the hunger pangs that, after several days in which they had failed to secure the food they needed, reminded them reproachfully of every occasion on which they had not followed sufficiently closely the lessons their mother had tried to impart.

Even now, even having survived their first orbit alone, bountiful spring yielding to barren summer and brutally cold winter, even as their hunting skills had sharpened and their success rate climbed, there were still periods during which that hunger gnawed away once more at their insides. And so, when they could not find seals or could not catch the ones they did find, they would do what they were about to do now, once they had caught the scent on the breeze, the unmistakable odor of a fresh kill and the enticing promise it carried with it.

They lifted their noses high into the air, their acute sense of smell establishing distance and direction, and they set off to find what they hoped still lay ahead of them. They padded swiftly enough to be at their destination as soon as possible, before other bears—perhaps much larger than themselves—descended upon the scene, as surely soon enough they would. But they dared not move so fast as to use up too much energy, and their approach required caution, too, for they did not dare to interrupt the meal that was still in progress.

Soon enough they found it, rounding a drift and seeing on the ice ahead of them the adult male, rounded and muscular, his heft and outline signaling clearly that he was a mature and successful hunter. They crouched down, retreated slowly and slightly, anxious but also wary. They would wait here, hopefully, and when he had taken his fill and left his kill, they would move in and gratefully help themselves to what he had left.

In her excellent book *Ice: The Nature, the History, and the Uses of an Astonishing Substance*, Mariana Gosnell lists a variety of words used by the Inupiat of northern Alaska to describe sea ice in its many forms, such as *sikuliagruaq* ("thick ice," defined as more than three feet or so thick), *agiuppak* ("a smooth wall of ice along the edge of landfast ice formed by other moving ice"), *aluksraq* ("young ice punched by seals forming a seal blowhole"), and *sagrat* ("ice floes of random sizes, beltways of ice with water on both sides"). Richard K. Nelson, in his book *Hunters of the Northern Ice*, adds such terms as *apuktak* ("ice coming together or hitting together"), *kaighechuk* ("rough ice"), and *napaiuk* ("one large piece of ice that has been washed up vertically to form a conspicuous landmark").

English can't quite keep pace with Inupiaq in terms of its nuanced definitions of sea ice, but neither is there a paucity of clarifiers, from frazil ice to slush ice, grease ice to pancake ice, ice floe, pack ice, fast ice, fresh ice, multiyear ice, and many others.

The reason that there exists such a multiplicity of descriptors for sea ice is that there is much to describe—not only the numerous forms that sea ice takes as it hardens, thickens, crashes, and splits apart, but also the way stations at which it pauses during its progression from liquid water to solid surface.

That journey begins, of course, as the temperature drops, steadily at first and then often precipitously, a sharp fall that presages the changes that lie ahead. The first identifiable sign that those changes are imminent is hard to define or even to describe other than as an apparent heaviness in the ocean. The sea seems sluggish, its swells dampened, as if the ocean is slowing down, resting in readiness for the transformation that is about to take place.

A few crystals make their appearance on the surface, soon swept together by the wind and the waves and knotting together to create a surface sheen that, as they increase in number, acquires a more oily appearance. This is grease ice, and it is a reliable harbinger of the onset of polar fall, which is itself but a brief layover on the road to winter.

The direction that grease ice takes in following that road varies according to the vagaries of the environment in which it forms. In predominantly calm waters, the grease ice forms long strips of interlacing fingers called nilas ice; in the more turbulent water of the open ocean, the constant movement breaks the continuous ice apart. The resultant fractured pieces are known as pancake ice but to at least one observer more closely resemble lily pads, approximately circular in shape and with raised edges from the grinding of pancake against pancake. The first pancakes are just a few inches in diameter, but as they collide they combine, absorbing each other and growing until they are several feet across.

At this stage, the ice remains sufficiently soft that a ship can push through it with relative ease, although the noise below decks, the grinding and crashing of ice against steel, suggests that greater resistance is being offered. A preponderance of pancakes should, however, be taken as a clue by any wary mariner that the time for departure for warmer climes is fast approaching.

A further slight drop in temperature, a marginal calming of the surface water, and now the pancakes are themselves connected by more grease ice, which itself solidifies further until firm floes begin to form. It is such ice floes that form the basis of the polar bear's realm. Compacted tightly together or separated by open water, it is these that the ice bear patrols in search of its prey.

A floe may be a few yards in size, or it may be as large as several miles. (Larger than five miles in length, an ice floe is commonly referred to as an ice field.) It is never static, either in location — pushed along by currents and the wind — or in size or shape. Over the course of its existence, a floe may break apart, shattered by collisions with other floes or cracking apart through inherent instability and the expansion of once-tiny cracks, time bombs that have sat deep within the floe from the moment of its birth. Or it may grow through the addition of other floes, two or more such masses grinding into each other, becoming intertwined, rising on top of and sliding underneath each other, the whole endeavor causing the floes to thicken and rise up into the air in the form of pressure ridges and to reach down deep into the water below. The scene is one of ongoing movement, a dynamic environment in constant flux.

This is the pack ice: at turns a stable, quiet, desolate, vibrant, crashing, thundering dervish of an environment. The pack constitutes the majority of polar sea ice, but not all of it. In the highest reaches of the Arctic, north of all landmasses, in the polar basin of the Arctic Ocean surrounding the North Pole and extending southward, the "polar pack" is, confusingly, a different beast. Here the ice has become more solid, even more twisted and misshapen; and whereas the ice of, say, Hudson and James bays or the southern Beaufort Sea freezes and melts on annual cycles, the ice of the true Arctic Ocean is just as likely to be several years old.

The reason behind this highlights another distinction between the Arctic and Antarctic. Because Antarctica is a continent surrounded by sea ice, during spring and summer the pack floats unencumbered into more temperate seas as it drifts apart, breaks up, and ultimately melts. But because the Arctic is, in contrast, an ocean surrounded by land, there is but one principal route — the Greenland Sea — via which floes can be transported into the warmer water that will inevitably spell their doom. Because of this, floes in the high latitudes of the Arctic Ocean are more likely to be "multiyear" ice: harder, thicker, and more concentrated than pack made up primarily of "first-year" ice.

But because it is floating freely, it is still pack, even if it is first-year pack's ugly and less frequently encountered uncle. In contrast, fast ice (known alternately as shore-fast, or land-fast, ice), which reaches outward from the shoreline to which it is attached, is spared the constant grinding and collisions of the pack and accordingly is flatter, calmer, seemingly more desolate — "like a lunar landscape," as Brendan Kelly describes it. Because there is less movement and turbulence, there are fewer cracks and openings, fewer opportunities for polar bears to find the prey they seek. And so, when bears head from land to their hunting grounds, they may only pause along the flat, featureless fast ice, taking advantage of whatever seal-like opportunities they may find en route, tarrying a while if pickings are good and multiple snowdrifts are hiding multiple young seals, before making for the pack.

Sea ice cover in the Arctic is never absolute. There may be large swaths where the surface appears to be nothing but varying shades of white, floes and fields of different shapes, sizes, and thicknesses, first-year blended with multiyear, and the whole twisted and bent into a visual cacophony of chaos. Persist, however, and eventually there will be an opening of sorts to the water beneath—a polynya, perhaps (an isolated patch of water surrounded by ice), a network of leads, a crack here, a hole there, perhaps even a stretch of open water.

When traveling through or across the ice, sea ice cover is measured in tenths as a means of discerning its extent and navigability. Ten tenths means the way ahead is completely full of ice, that there is nary a puddle to break up the solid mass, that forward motion via a seafaring vessel is not an option. Zero tenths means there is barely a chunk of ice to interrupt the journey, that there is nothing but open water in the immediate vicinity.

For polar bears, neither extreme is desirable.

Ten-tenths ice would be of no use to them at all, for although polar bears are creatures of the ice, the species on which they prey are only partially so. For those species, ice is a place of repose, not constant occupation, and one that they are able to reach only when there is access from their primary habitat. That habitat is the water between and beneath the ice floes, and although polar bears are technically marine mammals, in the water itself they are at a disadvantage to the animals on which they prey.

That is why polar bears seem to prefer ice cover that is less than 100, but more than 50, percent: sufficiently broken for them to be able to reach their prey animals, and sufficiently stable for them to be able to do so from the environment in which they have the upper hand. And that is why, for polar bears, unlike almost all temperate and tropical mammals, summer is a challenge and not a time of plenty; as the season advances and the Arctic warms, the ice melts, the cracks widen, and the environment shifts from one of sea ice dotted with fractures to one in which liquid water achieves equivalency and ultimately superiority over the frozen kind. In such a landscape, the location of briefly emergent seals becomes harder to predict before it happens, and more difficult to spot when it does. With seals surrounded by water and with only an increasingly remote and flimsy ice platform from which bears can launch themselves, stealth attacks become all but impossible and most other kinds of attacks unfruitful.

Only once has a polar bear been observed successfully hunting a seal in open water in the summer. In 1978, Don Furnell, a biologist from the Northwest Territories Wildlife Service in Canada, and David Oolooyuk, an Inuk hunter, were watching an adult male polar bear swimming in shallow water about seventy-five meters away from a ringed seal that appeared to be surfacing in approximately the same location each time it took a breath. The bear swam toward the seal's location, then abruptly stopped and lay motionless in the water. When the seal broke the surface a couple of feet away, the bear lunged, biting the seal in the back and killing it. Furnell and Oolooyuk speculated that the seal mistook the motionless bear for a floating piece of ice. Although the two men did note the presence of several seal carcasses along the beach, suggesting more than a few had been caught that way, the behavior has never been observed anywhere else, before or since. Perhaps that one bear deduced a successful method of aquatic seal hunting. But there is no evidence that the practice ever spread or even that that one bear ever repeated its trick.

But while polar bears eat primarily seals, and particularly ringed seals, they do not do so exclusively. They may, for example, either roam along the ice edge or wait motionless for an opportunity to seize a beluga whale. On occasion, belugas have become entrapped en masse in a polynya, unable to escape because the thickness and extent of ice cover in all directions would prevent them from coming up for air. That has sometimes prompted a polar bear feeding frenzy, one or more bears reaching into the water, grabbing a beluga, hauling it onto the ice, and then, before even stopping to eat the carcass of the whale it has just killed, doing the same thing over and over, seemingly unable to resist the primal urge to attack.

In presumably exceptional circumstances, bears will even leap onto the backs of passing belugas. It has not been seen often, but it has been seen: one bear was spotted killing two calves in this way in the space of twenty-four hours. It is a remarkable achievement, requiring the bear to not only make a perfect leap, but also debilitate and kill an entirely aquatic animal in the water.

Bears have even been observed attempting to chase down belugas by swimming after them—but in this they have not been seen to have any success, nor are the odds likely ever to be in favor of their doing so. In fact, researchers have seen belugas lunging at a would-be predatory polar bear and lashing at it with their tails, working as a group to chase it out of the water.

Polar bears, although perfectly competent and at home in an aquatic environment, are simply not credible predators when swimming. Belugas seem to know as much, judging from their apparent lack of fear of swimming bears, and the bears appear to be aware of it, too, particularly in the presence of walruses, which are far and away their most formidable natural foe.

"I've seen bears swimming in the water near walruses, and they're nervous bears," says Brendan Kelly. "They swim as rapidly as they can. In the water, they don't like being near them."

For good reason.

Inasmuch as walruses are known at all, it is, as Natalie Angier wrote in a delightful account of the species in the *New York Times* in 2008, "mostly for their sing-song linkage with a carpenter, an eggman and goo goo goo joob." But walruses are immensely social animals, crowding together on even a relatively small floe until there is barely any room for anybody to move. Males serenade females with a complex and fascinating repertoire of noises that Angier describes as sounding like "a circus, a construction site, a Road Runner cartoon":

> They whistle, beep, rasp, strum, bark and knock. They make bell tones, jackhammer drills, train-track clatters and the rubber-band boing! of Wile E. Coyote getting bonked on the head. They mix and match their boings, bells and knocks, they speed up and slow down, they vocalize underwater, in the air, at the bubbly border between. They sing nonstop for days at a time, and their songs can be heard up to 10 miles away. They listen to one another, take tips from one another and change their tune as time and taste require.

The whiskers, or vibrissae, on their muzzle are so sensitive and dexterous that, were one to place a piece of fish on them, a walrus could use them to maneuver that piece of fish toward and into its mouth. A walrus uses those same whiskers to grub around in the benthos, the ecosystem of the sea floor, looking for the clams and other bivalves on which it primarily subsists. Using immensely powerful suction, it is able to vacuum up as many as 7,000 clams a day. But at least some walruses yearn for larger prey.

On one occasion, Brendan Kelly was a part of a team studying whales in Arctic Alaska when a disturbance in the water caught observers' attention.

"The call came that there were two walruses in the water and they were fighting," he says. "They could see tusks, they could see blood flying and wild splashing, and they thought it was a fight between two walruses. So we eventually get out to this little polynya, and sure enough up pops this walrus, and we're crouched behind this little pressure ridge watching this walrus, and it swims around for a little while and then it dives down for about five minutes or so, about as long as you would expect it to take to dive down to the bottom and get some clams, and then he comes up and swims around for a little while. After a while we're thinking there was just one walrus and that the other folks were hallucinating, and then it did something I've never seen a walrus do before. It turns over on its back, like a sea otter, and it's swimming on its back, belly-up, and he takes one of his front flippers, and from under his armpit takes the hind flipper of a bearded seal, puts it between its two front flippers, puts the head of the femur to his lips, and . . ."

At this point Kelly mimicked the walrus's actions and, by way of illustration of what it was doing, broke the genteel silence in the hotel lobby in which we sat by making a loud sucking sound.

"It was sucking the meat off the bone of this thing," he elucidated. "It's really loud. Then he tucks it back in, swims around, dives back down, comes back up . . . he seemed to be alternating between clams and bearded seal. It was wild."

It was all too much for Kelly's dog, which had been straining at the bit to go chasing after this fascinating new animal until, unable to resist any longer, she ran to the edge of the ice. The walrus popped its head straight out of the water, focused its gaze on the interloper, and then sped toward it with apparently malicious intent, until Kelly emerged from behind his hiding place to prevent his canine from becoming another of the walrus dining options for the day.

When polar bears do hunt walruses, they do so from the relative safety of shore or ice floe, and they mostly focus their attention on calves. A Google video search will rapidly enough unearth footage of a polar bear attacking and killing a solitary bull walrus, but success is far from guaranteed. Hans Egede, who in the early eighteenth century was the first missionary to visit Greenland, wrote of walruses that they "are always in battle with the polar bear, and with their huge strong teeth, the walrus can make life hard to the polar bear; yes, the walrus often defeats the polar bear, or in the worst case they make a draw, so that they both end up lying dead at the spot."

In the 1930s, a hunter from Greenland saw a walrus take a preemptive, and fatal, strike against an unsuspecting polar bear.

"A polar bear had swum too close to an ice floe where a walrus family lay sleeping," he recounted. "But the bear was spotted and a large male walrus slid quietly into the water. It came up again shortly after behind the bear and slashed it with its tusks. They both dived and when they came up to the surface again, the walrus was gripping the bear with its front flippers whilst hacking at it with its tusks."

Although the bear escaped and reached shore, it soon collapsed and died, its skin perforated and organs gashed by the walrus tusks.

No surprise, then, that polar bears approach their sometime prey with a degree of caution that highlights their patience and underlines their wariness of the damage walruses can inflict.

One year, recalls Kelly, he and some other researchers "were on a ship, looking for walruses, counting them, studying them, doing exactly what we were doing on the *Arctic Sunrise* in 1998. There were twelve calves with twelve females all on one ice floe. We pulled up close, and we had finished counting them, and at that point the wind shifted slightly and they caught a whiff of us."

The walruses plunged into the water, the females doing what walrus mothers normally do: pushing their calves into the safety of the water first, and then following immediately behind. In the scrum, however, one mother was knocked off the floe before she had the chance to take care of her offspring, leaving her calf as the last remaining animal on the floe. Or so it seemed.

"At that very instant," continues Kelly, "from behind a pressure ridge, this bear stands up. It has clearly been there that whole time. It rises up, leaps over the pressure ridge, grabs the calf—doesn't kill it, picks it up like a kitten. It spins around, leaps back over the pressure ridge, jumps into the water, swims like crazy to another floe, runs across the floe, dives back into the water on the other side, gets back onto another floe . . . we followed him for thirteen minutes before he finally sets the calf down, puts his paw on it, looks around, and then reaches down and delivers the killing bite.

"Meanwhile, mom has risen in the water and she is bellowing like crazy, swimming round and round, making heart-rending bellows, calling and calling. It was very clear as soon as it grabbed the calf, that bear knew he had to get out of there, because he did not want to deal with mom. He really made good distance before he even took the time to kill the damn thing."

Most animals are to some degree risk-averse, their lives a constant cost/benefit analysis. A squirrel in the tree may see a pile of tasty nuts that has fallen to the ground; consuming them would alleviate those hunger pangs and provide a great deal of energy, but being on the ground might leave the squirrel far more vulnerable to a surprise attack from a fox or a bird of prey. Similarly, although they are unquestionably at the top of the Arctic pecking order, even polar bears are aware that they are not invulnerable. A walrus calf will yield a good many calories; it is relatively large and wrapped in a thick coat of desirable blubber. But an attempt to procure one may come at a terrible cost, exacted by a defensive and dangerous mother that may weigh two or three times as much as the would-be predator.

But while polar bears certainly have the option of seeking less dangerous prey, less dangerous does not necessarily translate to less resistant to capture. Polar bears prowl the pack, searching cracks and leads in the ice for seals that are coming up to breathe, but have no way of knowing in advance precisely where a seal might surface at any given time.

There do appear to be a few ways in which bears can improve their chances.

"If you go along a lead edge, there are often little peninsulas of ice that jut out into the edge, and bears typically seem to go out onto these peninsulas and wait," says Kelly. "And the seals seem to surface disproportionately near these peninsulas, and the bears will just snatch them right out of the water."

Even so, while life in the Arctic involves inherent uncertainty, a life spent sitting on ice floes and gazing longingly into open water would suggest one of positive irresponsibility. So polar bears diversify. They look not just for seals in the water, but for those that are resting on the ice along the edge. Hunting seals that have hauled out is in many ways an easier proposition: the seals stand out against the ice, and particularly during periods of pupping, there is an abundance of tasty morsels with neither sufficient awareness nor ability to escape on a consistent basis. Outside of that window, however, the seals that bask on the ice are the seals that, by and large, have survived long enough to be fully aware of the threats polar bears pose. Even when resting, they are wary and constantly on the lookout.

Were success dependent solely on being prescient enough to select the spot where a seal was about to come up for air or rest, or stealthy enough to catch by surprise any seal that had already done so, it is entirely possible that the ursine experiment in the Arctic might never have extended far beyond a few curious brown bears walking out onto the ice from time to time while looking to supplement their diet. Fortunately for our protagonists, there is one seal species that developed biological and behavioral traits that enabled it to survive and thrive in the Arctic for millions of years, but that also rendered it uniquely vulnerable to the monster that ambled over the horizon and shattered the predator-free calm of eons.

On the face of it, a ringed seal does not look very obviously more appealing to a polar bear than any other seal. At about five feet long and 150 or so pounds, an adult may be as little as 20 percent the weight of a bearded seal and be thus theoretically one-fifth as desirable. Pelage aside, it does not appear markedly dissimilar to most other seals, either. There is, however, one key difference.

At the end of its flippers, a ringed seal boasts five sharp, curved claws. These claws allowed the ringed seal to carve its own unique and lucrative niche in the Arctic marine ecosystem, for whereas other Arctic seals wishing to take a quick breath or a longer breather are obliged to make use of natural openings caused by the constant grinding, crashing, and splitting of ice floes, ringed seals can use their claws to create breathing holes of their very own.

"When the first molecule-thick layer of ice forms on the surface of the ocean, it's no big deal for a seal to push through that to breathe," says Kelly. "But when it gets many layers of molecules thick, when it gets inches thick, a bowhead whale or a walrus may be able to break through, but the small head of a ringed seal can't. So they reach up with these stout claws on their flippers, and they scratch a hole through the ice and poke their nose out to breathe."

As a ringed seal approaches a hole from beneath, it may pause, perhaps swim back and forth once or twice, and as it prepares to head upward and into the open air, it will blow a stream of bubbles ahead of it. The bubbles burst at the surface and in so doing, they clear away any detritus—bits of snow or ice—that may have accumulated. Ringed seals do this, Kelly suggests, not because such detritus blocks their passage but because it blocks their view, and a surfacing ringed seal is very keen indeed to try to ascertain whether anything is waiting for its emergence.

For, as much as the ability to create breathing holes has allowed ringed seals to flourish, it has also painted a target on their collective backs. A ringed seal will typically maintain no more than a half-dozen holes during the course of a season; in theory, all a bear need do is identify the location of a hole and be patient. And polar bears are nothing if not patient.

As fall progresses, snow accumulates on top of the holes, concealing them and insulating them. Ice continues to form, but with the snow's insulating warmth, it does so more slowly now than before. If the snow forms into a drift, a seal can, should it be so inclined, reach up into the snow and begin to carve out a snow cave. The cave functions as a place for the seal to haul out and rest on the ice, and while it may take advantage of the protective den for just a few minutes, it may also stay there for close to twenty-four hours.

Many of the caves, says Kelly, are quite beautiful inside.

"If you put your head in a snow cave and look up, there are these beautiful feathery crystals that hang down," he says. "It's frozen seal breath that's condensing on the ceiling and making these chandeliers, and the crystal light that's filtered through the snow is heavy in the blue spectrum, so the whole thing has a blue glow to it. It's just exquisitely gorgeous. I've looked at thousands of these things and every one of them is like a little chapel."

Some of them become not just chapels but maternity wards. In a fascinating echo of the polar bears' dens, it is in such caves, known as lairs—which, just like polar bear dens, may contain several chambers—that ringed seals give birth to and nurse their young.

Just like polar bears, ringed seal mothers must make their lairs in places where there is drifting snow. On the Arctic sea ice, that is almost invariably on the leeward side of pressure ridges formed by floes that have ground into each other, or where the pack has forced itself up against the fast ice and thrust small mountain ranges into the air.

But while polar bear mothers and cubs rest soundly in their dens, seemingly unconcerned by almost any disturbances that may pass by outside their cocoon, ringed seal mothers enjoy no such security. For patrolling the ice outside their snowdrifts, there is a very real threat.

The abundance of seal pups, and the inability of those pups to recognize or react to the danger posed by the prowling predators, makes springtime a time of plenty for polar bears, the time of year when food is both most plentiful and most vulnerable. (Their second most productive period, it seems, is early in the winter, when the rapidly freezing ice forces ringed seals to return to their breathing holes with great frequency to keep them open.)

The surface of the sea ice becomes a study in bear bacchanalia, polar bears feasting almost at will as they sniff out and crash through the lairs that have been painstakingly carved into drifts. It is a testament to the evolutionary and ecological success of ringed seals that, despite the massacre that is visited upon their young every spring, they continue to thrive.

The feeding frenzy is necessary; by the time the gorging is over, spring is well advanced and summer is fast approaching. The sea ice splinters, breaks apart, and melts; the watery leads become chasms in which seals can surface without fear of assault.

The bears are forced to retreat to ice that is thick or sheltered enough to persist during the warmer months, or even onto land, to await the time when temperatures drop once more, when the surface of the water freezes anew, and when they can once more wander, for mile upon uninterrupted mile, across the sea ice that is their unchallenged domain.

Life

Alone except for each other, the young polar bears face the most difficult period of their lives. If they survive to full adulthood, the age of five or six when they become sexually mature, their chances of surviving deep into old age are high; Ian Stirling has written that natural mortality in adult polar bears is so low that they are "virtually immortal." But the two and a half years until that time, after their mother has left them and before they have grown into full-size adults, are fraught with risk and challenge.

For two years they watched their mother, learned from her, and copied her every move — sometimes playfully, sometimes with more serious intent. But even as they began to hunt and kill seals, they did so with less frequency than she did and continued to rely on her milk for sustenance and certain survival. Now that option is no longer available to them.

They at least have the advantage of abundance, for the time of year in which they were left to fend for themselves was the same as when they first emerged from the den, when a profuse supply of young seals is theirs for the taking. But they must continue to refine their skills, and rapidly, in order to capitalize on this bounty; they must build up their fat reserves as extensively as they can in order to prepare themselves for the lean times that lie just a matter of weeks ahead. They must display the patience they lacked as occasionally dilettante hunters in their mother's company, expend far more time waiting for seals to emerge from their breathing holes, and not waste the opportunities that present themselves when seals do poke their heads out of the safety of the frigid water below. In the harsh Arctic environment, there is no margin for error.

If dogs dream of chasing slow-moving rabbits through grassy fields on a sunny day, then for a wishful, dozing polar bear, paradise is surely a place in which sea ice cover is plentiful and solid, but patterned with cracks large enough for a lazy seal to poke its head through and linger invitingly.

When it rouses from its reverie, the dozing bear shakes off the snow that has settled gently on it during its slumber and rises slowly to its feet. It sniffs the air, blinks softly, and then, after a final shake of its head, it walks.

The Inuit of northwest Greenland call the polar bear *pisugtooq*, "the wanderer," for that is what it does. When it is not sleeping—and polar bears, like humans, sleep for approximately eight hours out of every twenty-four—it is walking. Its stride is languid, apparently without aim or intent; but it is alert at all times, and the moment it sees or smells a seal on the ice—often from a distance of several hundred yards—its demeanor changes in an instant.

It immediately stops, stands perfectly still, and waits, staring at the seal, sometimes for several minutes, as if sizing up the distance to the prey. Then, slowly but surely, it begins its stealthy march. As it moves forward, the bear lowers its head; once it draws within striking range, it sinks to a semi-crouch.

That polar bears are found throughout the Arctic but nowhere in the Antarctic is evident in the behavior of seals at both ends of the Earth. In the Antarctic, seals that have hauled out onto the ice observe approaching explorers or researchers, whether on foot or ship-bound, with mild curiosity but little if any alarm. Save for the restrictions imposed on such behavior by Antarctic Treaty regulations, it would be quite possible to walk up to a basking seal and not elicit much or even any alarm until close enough to reach out and touch it—and often, not even then. For there are no land predators in Antarctica, and a pinniped on the ice is in danger only if it has chosen to repose on a floe that might be tipped over by an observant killer whale. In contrast, ringed seals in the Arctic are bundles of restless energy, napping for perhaps a minute and then lifting their heads to scan their surroundings for several seconds, their large eyes on the alert for any movement, their sharp hearing tuned to detect anything untoward. As soon as a seal tenses in response to the sight or sound of movement, or merely raises itself onto its front flippers for a precautionary inspection of its immediate environment, the bear freezes in its tracks, able to hold its position without a wobble or a twitch, until the object of its attention, apparently unable to pick out the whitish mass of the approaching menace from the background or detect the gentle padding of its paws on the ice, lays down its head once more.

Slowly, surely, the bear closes in on its target. Then, suddenly, it charges.

Polar bears are not built for sustained speed; their heft and insulation work against them and they soon begin to overheat. But over distances of forty or fifty feet, they are capable of shockingly explosive bursts, and the moment they close to within that distance of their target, they take off, pounding forward with no further pretense of subtlety or subterfuge. One can only imagine the overwhelming shock and panic that must crash like a tsunami over the seal as this monstrous beast appears as if from nowhere, casting aside its cloak of invisibility and roaring toward it with frightening resolve and force.

Desperately, the seal flops toward the nearby breathing hole through which it emerged onto the ice, and its frantic bid to seek the sanctuary of the water that must suddenly seem so far away is an illustration of why ringed seals predominantly haul out of the ice by themselves and not in groups of several: a mass of blubbery bodies flopping pell-mell for safety would obstruct one another's passage so that each and every one of them would be at the bear's mercy.

(Indeed, notes University of Alaska ringed seal expert Brendan Kelly, those occasions when numerous seals do haul out by one hole frequently have entirely predictable outcomes. "When such groups are alarmed, they all try to escape down the hole at once," he says. "Most often that results in two or three seals with their heads submerged but their bodies unable to squeeze in at the same time. It is almost comical to see their hind flippers flapping madly in the air.")

As it is, by the time the seal reacts to the thundering onslaught, it is too late. The bear's jaws clamp shut around its head, then again, then again, before the predator hauls the carcass away.

A stalking polar bear does not always approach across the surface of a floe. On occasion, after staring at the intended victim as if to memorize the route, a bear will quietly slip into the water and swim through breaks in the ice, surfacing to breathe so stealthily, writes Ian Stirling, "that only the tip of the nose breaks the water between dives as the bear moves closer to the seal." Finally, Stirling continues, "it gets to the last available breathing hole before reaching the seal and slips out of sight again. After an eternity of suspense-filled seconds, the water in front of the seal explodes as the bear suddenly claws its way onto the ice after its prey."

While that may be the most spectacular mode of attack, it is not necessarily the most effective. More often than not, Stirling notes, the seal manages to evade the bear's claws and slip into the water, some-times just inches ahead of the pursuing predator. More successful is a variation on the theme, in which bears approach their prey by means of the water-filled channels that form on top of melting sea ice in the summer, slithering along almost completely out of sight until close enough to strike.

One bear that used that particular technique showed such intelligence and awareness that even Stirling, who has studied polar bears in the wild for more than thirty years, was taken aback. After spotting a seal lying on the ice, he records, the bear stopped in its tracks and stared intently, without moving, for several minutes at a series of channels in the ice that led toward its potential prey. Then it quietly slipped into one of them and paddled gently ahead. So far, so unexceptional; what stunned Stirling was what happened when the bear came to a fork in the channel. Initially, the bear turned right at the fork; but after proceeding "about a body length," it stopped and, without lifting its head, backed up and then took the left channel, which was the one that led to the seal.

"I was amazed," Stirling wrote. "I have seen several bears lift their heads slightly to check their bearings but I have never seen such a demonstration of conscious memory."

Polar bears are more likely to use stalking as their means of predation in summer, when the ice is melting and narrow leads have widened to gaping channels, when the seals that haul out to rest do so in the open air. In fall and winter, those same seals take advantage of the breathing holes that they maintain in the ice, which are nominally hidden and protected by the seemingly anonymous snowdrifts that form over them. But a ringed seal's sanctuary is one that a polar bear's powerful sense of smell can pinpoint literally from miles away.

A polar bear zeroes in on a seal sheltering within its lair in the same way it does on one that is basking in plain view: its apparently aimless wandering comes to an abrupt halt, its nose sniffs the air. The bear may clamber onto a pressure ridge to gain a better sweep of the area, may move its head from side to side as if triangulating the scent and confirming the direction. Then it sets off toward the hidden bounty until, as it closes in on the lair in question, it suddenly speeds up and then, at the last moment, rears up on its hind legs and drives its forepaws down with a crash through the roof.

"A lair might be a meter, could be five meters long, and might have multiple chambers, but somehow, from that point on a pressure ridge, a bear is able to figure out not only which snowdrift to go for but also at which point in a snowdrift to make a hit," marvels Brendan Kelly.

If the bear is lucky, its attack crashes the roof onto the breathing hole itself, blocking the exit and trapping any seal that had been lying in the snow cave. Sometimes, however, the seal slides through the hole and into the water in advance of the bear's final lunge, or had already recently vacated its rest area.

When that happens, the bear will wait.

At times a bear must wait very patiently indeed, for perhaps an hour or even more. Most of the time, it will lie quietly on its stomach and chest, its chin close to the breathing hole or ice edge. It must remain not only still but completely silent: even the slightest movement on the ice is magnified as it reverberates into the water below. Such is the patience that some of the first explorers and naturalists to observe the behavior assumed that the bears were sleeping on the ice.

Bears do not, contrary to popular myth, cover their black noses with their paws to make themselves completely white. An extension of that story posits that it is always the right paw that covers the nose, and that all bears are left-handed — a wrinkle that also, from hours of observation of bears swiping seals with either forelimb, appears to be inaccurate.

Another tale, yet to be disproved, is that bears will, at least on occasion, be inventive: that they will, for example, build walls of snow behind which to hide as they wait at a breathing hole, or even that they will push blocks of ice ahead of them as they creep forward during a stalk. Although both prompt the raising of an eyebrow, the latter seems especially improbable; seals are as attuned to the aural as to the visual, and the sound of ice scraping against ice would be far more likely to carry into the water and the ears of an attentive seal than the gentle padding of a predator's paws. (It is more likely that bears may push snow ahead of them with their noses as they crawl; polar explorer Eric Larsen recalls that a bear that apparently was stalking him on the ice of the Arctic Ocean in 2005 was doing just that.)

There is, however, some circumstantial evidence for another claim, that bears will sometimes use tools — specifically rocks or blocks of ice — to break through the most resistant of snow caves. Bears in captivity have certainly been observed to throw pieces of ice, and Erik W. Born relates this tale from Inuit hunter Kristian Eipe, who encountered the body of a recently killed walrus: "Before the bear had sat down to wait for a walrus to come up for air, it had fetched a lump of sea ice from the tidal zone," Eipe asserted. "It had prepared the lump of ice so that it was completely smooth, thereby making a tool which it could use to smash the walrus on the head . . . The bear's claws had made deep marks in the fresh ice. It had attacked and hit the walrus on the head with its weapon . . . The bear had crushed its skull from the muzzle to the back of the head, the skin there was torn to pieces, such was the force of the blow." In 1972, H. P. L. Kiliaan of the Canadian Wildlife Service was sledding across Sverdrup Inlet on Devon Island with two Inuit when one remarked he had seen a place where a bear had used a block of ice to smash a ringed seal lair. The three went to investigate and, reported Kiliaan, found a lair that had been broken open, the tracks of a female bear and two cubs, and, next to the breathing hole, a chunk of freshwater ice he estimated to weigh about forty-five pounds. By following the tracks, Kiliaan deduced that the bears had broken the ice from a much larger piece and then rolled it toward the breathing hole. He wrote:

What happened next is not certain, but we may consider the following possibilities:

1. The bear used the piece of ice to smash through part of the snow roof. Once the roof was thinner it would be possible for the bear, using its own weight, to break through the rest. Then should the seal surface, the bear would be able to take it by surprise.

2. The bear heard the seal come to the surface and tried to smash it on the head with a piece of ice, in an attempt to stun or kill it.

3. The bear, after an unsuccessful attempt to catch the seal in the conventional way, broke off the piece of ice in "anger," frustration, or play, and rolled it towards the opened aglu,* and by sheer coincidence left it there.

That bears do lose their temper or vent their frustration on those occasions when they fail to make a kill — when a seal escapes a bear's lunge, effectively wasting the minutes or even hours the putative predator has devoted to securing a meal — is another tale told often and mostly apocryphally. In his book *The World of the Polar Bear*, Richard Perry cites several observers who report having seen flashes of bear temper in the wake of an escaped seal, ranging from splashing the water surface furiously, to swiping at snow and throwing it in the air, to even hitting a nearby rock outcrop and, so the unsubstantiated story goes, supposedly breaking the bones in its paw in the process.

But when a bear does succeed, when it surprises a seal and leaps in time to seize its victim in its jaws, when it is able to rip its prey away from the sanctuary of its breathing hole, it is an expression of such raw power that an observer may feel chilled by the ferocity of what he or she has just witnessed. Charles Feazel, a polar geologist who in the course of his work had many encounters with polar bears, is a case in point, as detailed in his book *White Bear*:

---

* An Inuit word for a ringed seal breathing hole, often used in popular terminology.

Yesterday I watched a bear kill a seal. I shouldn't have. Like the strength sapping cold, the memory seeps inward, displacing all focus on my scientific mission. The scene was a grim lesson in arctic efficiency. Shuffling along through the snow, the big she-bear looked peaceable enough. Then scenting, through more than two feet of snow cover, a seal's breathing hole in the ice, she froze . . . Suddenly, rearing up on her hind legs, she towered, motionless, a silent, menacing apparition almost eight feet tall. She waited. Then, with a dive so fast the eye couldn't follow, she plunged nose first into the snow. A great cloud of white powder exploded into the still air, mercifully obscuring the seal's final agonies. With her massive jaw and thick neck muscles, the bear crushed the seal's skull and lifted its 150-pound body clear of the water. The power of the upward jerk pulled her prey through the narrow opening in the ice, and broke most of the seal's bones. Swinging her long, snakelike neck from side to side, the bear flayed the seal . . . The bloody carcass, sleek as a red torpedo, slithered across the snow, a grotesque puck in a deadly game.

A bear's principal nutritional target is not the seals' meat.* The proteins in red meat are difficult to break down, particularly in an environment where liquid fresh water—essential for protein absorption—can, at least at some times of the year, be difficult to find. Instead, the bears zero in on the seals' thick layers of fat, which are rich in easily digestible calories. So focused are polar bears on seals' fat that researcher Andrew Derocher has referred to them not as carnivores, but lipovores.

(They are not alone in their lipophilia. One day on the *Arctic Sunrise*, as I stood on deck simultaneously savoring the view and detesting the cold, I described to one of the scientists on board my impression of

---

* Most of the time, anyway. When first out of their dens, the bears eat muscle as well as skin, blubber, blood, and organs. At that time of year, notes Brendan Kelly, "I often find only the tympanic bullae [ear bones] left at a kill site." As the bears fatten, they become more selective and later in the summer take only the blubber.

muktuk. Muktuk is the name given to the skin and blubber of bow-head whales, often thinly sliced; during a visit to an Inupiat community earlier in the voyage, I had been offered some and had accepted. To coastal peoples of the Arctic, *muktuk* is a delicacy; to my admittedly vegetarian palate, it was chewy, oily, and indigestible. What I was missing, said the scientist—who had spent much time on the ice with Alaska Natives—is that when the temperature is firmly in the double digits below zero, the body craves nothing so much as fat, which is easily digestible and readily releases warming energy. In that context, the idea of whale blubber as a satisfying source of calories makes perfect sense.)

After killing a seal, a bear tears chunks away from the corpse and swallows them almost without chewing, desperate to consume as much as possible as soon as possible before the scent of the kill attracts competitors. As the meal progresses, the bear becomes more refined, almost to the point of an incongruous daintiness, using its incisors to shear away the fat from the meat and, if it has the time, peeling the apparently undesirable skin from the fat.

Polar bears are astonishingly efficient predators, not just in their ability to catch prey but also in their digestion of that prey. A 1985 study in the *Canadian Journal of Zoology* ca lculated that they can assimilate 97 percent of the fat they consume. The report's author, Robin Best of the University of Guelph, calculated that the average active adult polar bear would need roughly five pounds of seal fat a day to survive—a target easily reached with any seal more than a month old and comfortably surpassed with an adult bearded seal, which on average would provide enough sustenance for a solid week.

If uninterrupted, a bear may feed for an hour or more. If the kill is small or the bear is large (or especially hungry), there may be little left at the end of the meal; more often, however, there is some fat and much meat remaining. Unlike brown bears, polar bears do not cache the remains for future feeds; they may make rudimentary, instinctive scratching motions in the snow, but generally they leave the remains of the carcass on the ice, where they may swiftly be fallen upon by arctic foxes or subadult polar bears, which frequently take advantage of their elders' leftovers to sustain them as they learn to hunt.

The meal completed, a polar bear cleans itself. In summer, a thirty-minute meal might be followed by a fifteen- or twenty-minute wash, the bear rinsing and licking its paws and muzzle in a pool of water; in winter, in the absence of liquid water, a bear will rub its head in the snow and roll on its back, doing so repeatedly until the blood, oil, and blubber of its victim are cleansed from its fur and face. Polar bears observe cleanliness to the point of fastidiousness; Richard C. Davids has written that he has "seen them back up to the edge of the ice to defecate in the water."

Having feasted and having cleansed, the bear may pause a while, sniff the air, look around, and amble away. Chances are it will wander up a slope, find a spot sheltered from the wind, make a depression in the snow, and clamber in. And then, as the drifting snow gently covers it up again, the bear will do what many a human will do after a hearty meal: it will curl up and go to sleep.

As the adult bear wanders off in search of repose, our two subadults move in to take advantage of what has been left behind. As they refine their own hunting and survival techniques, they sustain themselves by scavenging the remains of kills made by larger, more experienced bears. They are not exactly starving, not particularly skinny, but they have not broadened out into the healthy, rotund profile of their elders. As Ian Stirling has written, although he has encountered many healthy subadults, he has not come across many fat ones. In those few areas where polar bears and humans overlap, it is almost invariably subadults that become problem bears, skulking around refuse dumps in search of extra nourishment.

Polar bears may take advantage not just of human waste, but of human hunting: in the Beaufort Sea region, and particularly near the village of Kaktovik, dozens of polar bears each year have developed the habit of gathering at the butchering sites of bowhead whales that are killed by Inupiat whalers. "The value of this alternate food is apparently great," notes Steven Amstrup, "as nearly every bear seen near whale carcasses in autumn is obese."*

But such riches are not available to our two young bears; were they to attempt to partake in such a feast, they would soon be bullied and crowded out of the reckoning by more established bears. Besides, the issue is moot; there is no whaling in Hudson Bay, the frozen surface of which they now prowl. They must instead depend on their own wits and instincts, must build up as much energy and fat deposits as they can during the abundance of spring, to see them through the lean times of summer. And as they learn to hunt, feasting on the scraps from others' tables will help them to do so.

On this occasion they are fortunate. They were the nearest bears to the kill, and their keen noses picked up the scent of the dead seal before any others. Uncertain of their status, not confident in their strength, they did not approach until the bear that had killed the seal had finished and moved away; only when they felt the scene was safe did they move in for the scraps.

An adult bear might not have been so circumspect. The reason why a bear—at the risk of mixing mammalian metaphors—wolfs down its food is because of the urgency of eating as much as possible before a rival, guided by its tremendously powerful sense of smell, zeroes in on

---

* For individual bears on a short-term basis, anyway. The amount and availability of the extra food source are not enough to provide any kind of boost on a population level. Plus, as Geoff York, formerly of the USGS and now with the World Wildlife Fund, points out, "it is essentially a subsistence 'dump' scenario, not sustainable or natural, and will end badly for people and bears alike if allowed to persist as currently practiced in Kaktovik." For this reason, the larger settlement of Barrow has a program to dispose of the waste from its fall whaling season and also actively seeks to keep bears out of the community.

it. In response to the intrusion, the first bear will likely break off from its booty to snarl and threaten the new arrival, but it is rare that two or more bears will come to blows over the carcass of one seal; the chance of one or both animals suffering severe injuries as a result of any fight is too great, and the potential reward—a part of one seal carcass—is far from sufficient to justify such a risk. It is more likely that after a period of posturing and growling, the rivals will reach an uneasy truce and share what remains before going their separate ways.

Younger, smaller, and less experienced bears are another matter. They have neither the strength nor the confidence to resist any challenges and are frequently driven away from their kills by older, larger, more aggressive animals (and occasionally, although seemingly rarely, may even be killed for their impudence). Even now, as the two subadults chew at the carcass that has been left behind, absorbed in extracting as much nourishment as possible from the remains, they are on the alert for competitors. The wind picks up and the snow begins to swirl, the harsh environment a stark contrast to their formative days in the warmth of the den. Then, life was simple and secure. Now, the world is far more uncertain and intimidating.

Every individual polar bear has what is known as a home range, but such ranges are quite distinct from territories in the more widely understood sense. Whereas brown or black bears may find a patch of land ripe with berries and salmon, claim it, and defend it fiercely, polar bears have no such luxury. Ecological and environmental conditions in the Arctic are unpredictable; the desirability and productivity of one area can change from year to year, month to month, and even week to week. And all the while, the surface on which they walk is constantly moving, shifting beneath them—"Imagine," says polar bear researcher Geoff York, "living on a treadmill."

While a home range could encompass a great many leads in the ice one year and thus require a bear to cover only a relatively small area, by the following year ice patterns may be completely different. Even in those areas where ice does not melt completely in summer, there will be significant changes from one season to the next. Polar bears' environment is so dynamic that they simply cannot afford the luxury of staking out their own patches, even if those patches weren't constantly moving beneath their paws. Bears go where the climate, environment, and ecology are to their advantage; a bear that has a vast expanse of solid, seal-free ice to itself may be lord of its own substantial domain, but it will soon enough be dead. As a result, polar bears congregate in the most advantageous areas, their home ranges overlapping or even overlaying each other.

That polar bears have home ranges at all is a relatively recent discovery, a consequence primarily of studies involving satellite tracking of bears fitted with radio collars.* It was long thought that they roamed across the Arctic sea ice, every bear a member of one giant family that knew no limits or boundaries. In fact, not only does each bear largely confine itself to a general range, but the entire global population of polar bears may be divided into a number of discrete subpopulations. Traditionally, polar bears have for management purposes been divided into nineteen subpopulations, thirteen of them in Canada—nine in the central Canadian Arctic, from Hudson Bay in the south to Kane Basin in the north; northern and southern subpopulations in the Beaufort Sea, off western Canada and northern Alaska; and one each in Baffin Bay and the Davis Strait to the east. Delineation of those subpopulations was largely based on bears' fidelity to denning and foraging areas and summer refuges such as the permafrost dens used by bears from Hudson Bay.

---

* Radio collars are fitted only on female bears. The thick necks of males are wider than the bears' heads, creating a kind of funnel shape from shoulders to nose, meaning any collars would slide right off. The necks of females are narrower, and as a result, collars can remain affixed until, at a preprogrammed time, they unlock and fall off. However, new developments are making it possible to monitor males' wanderings with small radio receivers that attach to the bears' ears.

But in a 2008 study in the journal *Oryx*, Gregory Thiemann, Andrew Derocher, and Ian Stirling took a closer look and examined Canada's polar bear populations for genetic, geographical, morphological, and ecological similarities and differences. They noted that, for example, Beaufort Sea polar bears tend to be smaller and to reach maturity later than those farther east, and that Banks and Victoria islands act as a barrier between bears from the Beaufort and bears from the central Canadian Arctic. Although the islands of the Canadian Arctic Archipelago serve to separate the bears of that area into six subpopulations, as had long been recognized, Thiemann, Derocher, and Stirling concluded that the separations were relatively weak and the genetic similarities comparatively strong. In contrast, the rugged topography and ice fields of Ellesmere and Devon islands and the thick, relatively seal-free, multiyear sea ice between the Archipelago islands effectively separate the central Canadian Arctic bears from those farther to the north. Contrasting ocean current patterns in Baffin Bay and neighboring Davis Strait mean that, for example, the sea freezes up earlier and breaks up later in the former than in the latter, creating different ecological conditions that have led to genetic differences between the two subpopulations. And the three previously recognized subpopulations of western Hudson Bay, southern Hudson Bay, and Foxe Basin in fact constitute one genetically distinct unit.

Thiemann, Derocher, and Stirling ultimately concluded that there are in fact five "designatable units" of polar bears in Canada—in the Beaufort Sea, High Arctic, Central Arctic, Davis Strait, and Hudson Bay. Among them, these five units likely contain approximately 60 percent of the world's polar bears.

The 40 percent of non-Canadian polar bears are for now considered to belong to six subpopulations: in East Greenland; the Barents, Kara, Laptev, and Chukchi seas; and, most northerly of all, the Arctic Ocean.

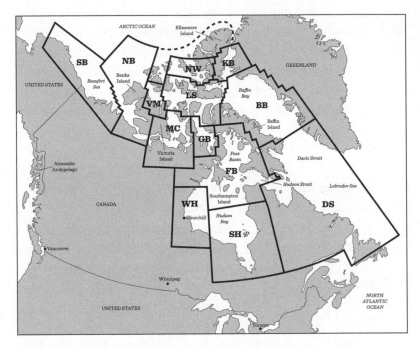

NB Northern Beaufort Sea; **SB** Southern Beaufort Sea; **VM** Viscount Melville Sound; **LS** Lancaster Sound; **MC** M'Clintock Channel; **GB** Gulf of Boothia; **NW** Norwegian Bay; **KB** Kane Basin; **BB** Baffin Bay; **FB** Foxe Basin; **WH** Western Hudson Bay; **SH** Southern Hudson Bay; **DS** Davis Strait

NB Northern Beaufort Sea; **SB** Southern Beaufort Sea; **VM** Viscount Melville Sound; **LS** Lancaster Sound; **MC** M'Clintock Channel; **GB** Gulf of Boothia; **NW** Norwegian Bay; **KB** Kane Basin; **BB** Baffin Bay; **FB** Foxe Basin; **WH** Western Hudson Bay; **SH** Southern Hudson Bay; **DS** Davis Strait; **AB** Arctic Basin; **EG** East Greenland; **BS** Barents Sea; **KS** Kara Sea; **LPS** Lapteov Sea; **CS** Chukchi Sea

The ice in the high latitudes of the Arctic Ocean is generally, although not constantly, thicker and less fissured than that near the coasts, and seals are more disparate and harder to encounter. Not without reason is the region decried as the "polar desert." Ralph Plaisted, who in 1968 became the first person incontrovertibly to reach the North Pole, remarked of the area that it was "absolutely desolate." But, he remarked, "we saw fox tracks every day. The foxes follow the polar bears to feed on carrion they leave behind." What he could not understand, however, was the fact that, for all the tracks he spied, he "never saw a fox or one single bear."

For some considerable time, there was uncertainty over whether polar bears seen in the Arctic Ocean basin were truly part of a subpopulation or whether they were merely visitors, extending their range temporarily in search of food. But some of those bears would have to be extending their range very far.

In October 2003, the nuclear-powered submarine USS *Honolulu* broke through the ice 280 miles from the North Pole, where it was promptly investigated by a female bear and two adolescent cubs. At least those bears only looked. Six months earlier, an officer on board the USS *Connecticut* began scanning the immediate area after that submarine also surfaced through the pack ice, only to see a polar bear stalking the vessel and then chewing on and swatting the tail rudder. Damage to the sub was reported to be minor: "Rear rudders of U.S. submarines aren't designed as snacks," said a naval officer.

During their 2005 attempt to cross the Arctic Ocean in summer, Eric Larsen and Lonnie Dupre were bedeviled by the attentions of polar bears in the broken coastal pack off Siberia. During a second attempt the following year, they saw no bears at all—until, amazingly, they were at the North Pole.

"I can't even imagine that. We saw a bear at the North Pole," marvels Larsen. "At the freaking North Pole. Unreal."

• • •

Polar bears travel immense distances. In the Beaufort Sea, where polar bears have been studied by radiotelemetry for over twenty years, it is not uncommon to measure a bear traveling at between two and three miles an hour for several hours, fast enough to make good headway, not so fast as to prompt overheating, or covering thirty miles during a day. Over the course of a year, Beaufort Sea bears have logged, on average, a little under 2,200 miles and at most 3,800 miles during their wanderings from one weekly location to another. And although they generally maintain home ranges, those ranges can be vast. In the Beaufort Sea, the average "annual activity area" for studied bears was 57,500 square miles, the smallest being 5,000 square miles and the largest, 230,500. The annual activity area is not necessarily synonymous with the home range, and may in fact be just a portion of it, as changing ice conditions frequently prevent bears from using the entirety of their ranges any given year and frequently force them to use different areas from year to year. In the Chukchi Sea, six bears studied using radiotelemetry covered on average approximately 95,000 square miles annually; other studies have shown similarly large areas in the Davis Strait and Baffin Bay. In contrast, bears in the interior Canadian Arctic Archipelago generally cover an area each year that, while still substantial—up to almost 9,000 square miles according to one study—is significantly less than that of their aforementioned counterparts.

Close examination of seasonal bear movements in the Beaufort Sea helps explain the difference. Here, the periods when polar bears cover the greatest area are June to July and November to December—the times when ice is melting during the spring and re-forming during the fall. During spring, previously reliable leads widen and ultimately disappear, and during fall, many have yet to truly form. The ice is patchy and unpredictable, and fertile hunting ground has yet to reveal itself; accordingly, bears must travel far and wide in search of seals. Stronger currents in the more wide-open spaces of the Beaufort and

Chukchi seas and Baffin Bay help explain the greater volatility of ice conditions and the likely need for bears to not only range more widely but also adapt on the fly to ever-changing ice conditions.

Such volatility does, however, have its advantages. As the ice moves and grinds, it creates openings that allow more sunlight to pass through into the water below, helping fuel the growth of the plant and animal organisms that form the basis of the marine food web. Not only does that help nurture an ecosystem in which seals may be relatively abundant, but the fractures in the ice allow easier access to air for those seals to breathe.

Conversely, in Viscount Melville Sound in the Canadian Arctic Archipelago, polar bears in general showed less movement on a yearly basis, presumably because most seals in that area are concentrated along tidal cracks and pressure ridges that parallel the coastlines of the area's islands. The ice here is thicker and more stable, with a greater percentage of solid multiyear ice, and the seals' whereabouts are more predictable — but, largely because of that solid ice cover, their abundance is reduced and foraging conditions are relatively poor.

The fundamental centrality of sea ice to the polar bear's existence is underlined by tracking radio-collared bears and layering their movements on that of the ice over the course of a typical year. Between May and August, as the ice of the southern Beaufort Sea is degrading, bears in that subpopulation not only expand their activity areas, but they move north, where the ice pack remains robust; their movements are extensive and rapid, as they seek out feeding opportunities. In October, as the first pack starts to form over shallow water near shore, they swiftly head south and disperse east and west along the coasts as the ice solidifies over the course of winter.

As the bears move, so does the ice beneath their feet. In the Beaufort Sea, it travels in an immense gyre, carried by the currents south along the west coast of Banks Island, west along the Canadian and Alaskan coasts to a point just past Point Barrow, after which it heads north once more. And yet, despite this constant movement, despite the lack of what, to us, would be obvious visual cues in a landscape composed of solid pack stretching to the horizon, polar bears are able to compensate. Ian Stirling notes that he has often recaptured polar bears in the Beaufort Sea, "far from sight of the nearest land, within a few kilometers of where they were first caught two, three, or even more than ten years earlier."

Arguably even more impressive are those bears that den on the pack ice; for several months, as the cubs suckle from their mother in the dark warmth of their den, currents transport them hundreds of miles from their original point. No other vertebrate is transported passively and blindly as far and wide as are denning polar bears. And yet, somehow, even though when they emerge it is in a place where they have never been and where they did not choose to travel, they are able to find their way back to the area from which they came, and which the mother had already established as her home range.

Exactly how polar bears orientate and navigate in this way is unknown. Perhaps a clue lies in the discovery by researchers Malcolm Ramsay and Dennis Andriashek of the route taken by female polar bears with cubs as they left their maternity denning area south of Churchill in February and March and headed back onto the ice. The shortest route would have been due east. Instead, they took straight courses that paralleled each other, the mean course of which was 39 degrees True (a little north of northeast). That might suggest that, like migrating birds or, it is believed, whales and dolphins, polar bears take advantage of cues in the Earth's magnetic field, that they possess a kind of mental compass that leads them, instinctively and unerringly, to where they need to go and where they have been before.

Polar bears are at once solitary and social. They are solitary in the sense that they do not, outside of mothers with cubs, live in any kind of organized familial or hierarchical structure. They do not live in pods, as do whales, or in packs like wolves, or in prides like lions. And yet, in some ways, they are as social as — or even more so than — all of those.

Like tigers and leopards, polar bears hunt alone, make no lasting connections with their mates — do not, in fact, see them after copulation has been completed — and, in the case of males, have nothing to do with their offspring. And yet, largely because of the particular circumstances of their environment and its challenges, they frequently come together in what can only be described as social settings and display what can only be considered social behavior.

As Richard C. Davids has observed, "Polar bears used to be considered lonely nomads, among the most solitary creatures in existence. I wondered about that on my first trip north, when I saw five big males lying in the kelp along shore, their noses all but touching."

Early during his research on Wrangel and Herald islands in the Russian Arctic, Nikita Ovsyanikov noticed a variety of social interactions that caused him to ask whether the polar bear is "a solitary ice wanderer or a more social animal?" The answer, he found, was more complex than prevailing theories of the bears' purely solitary nature initially supposed.

Ovsyanikov was especially well positioned to make those observations. The hostile, uncertain nature of the Arctic environment means that, for all that polar bears mark out their own clearly defined home ranges, bears — like all animals — must go where the food is. Perfect hunting conditions, high prey concentrations, or a sudden absence or abundance of sea ice may all lead on occasion to a polar bear agglomeration. Given that the fickle nature of sea ice makes defending territory a proposition of dubious value, and the considerable size and strength of other polar bears make it one of particular danger, a certain acceptance of others would therefore be explicable. Indeed, Ovsyanikov at times counted as many as 140 bears gathered in close proximity at his observation area at Cape Blossom on Wrangel Island. As Geoff York explained it to me, "It isn't that polar bears are asocial. It's that they have no particular reason to be social most of the year. When it suits them, they are quite gregarious."

There were occasions during Ovsyanikov's research when bears showed suspicion of, and even hostility toward, each other, mostly when females with cubs resisted approaches toward their kills by adult and subadult males. There were, however, numerous other instances when bears shared food willingly.

"In fact," Ovsyanikov wrote, "when eating from the same carcass, the polar bears I observed were often more tolerant of each other than highly social wolf packs are around a kill. I soon realized there was more to it than simple tolerance."

Ovsyanikov noticed that bears seemed to take turns in feasting on, even "possessing," the carcass, but that when others approached, perhaps circling around the carcass, gently initiating nose-to-nose contact, then, once they had finished their meal, the original bears willingly gave up their place at the table.

It is not atypical for a bear that has killed a seal to allow, after some growling and lunging, a rival to partake in its kill. But the extra degree of sharing documented by Ovsyanikov might be at least partly explained by the fact that the bears he was observing were feasting not on ringed or bearded seals, but on walruses.

As we have already seen, hunting walruses is an altogether different proposition for polar bears than stalking or waiting for seals. More than two thousand pounds of meat and blubber wrapped in thick hide and punctuated by two curling tusks, walruses are fearsome opponents. A desperately hungry bear may leap on the back of a walrus, biting its shoulders and neck and attempting to bring it down, but the would-be prey often succeeds in dragging its putative predator into the sea, where the pursuit will end; although polar bears are marine mammals, in the water they are by far no match for the more completely aquatic pinnipeds. From Ovsyanikov's observations, "only a few experienced animals were able to make a kill, and the walruses killed were all calves."

Notwithstanding their size and strength, walruses seek safety and security in numbers, frequently gathering in large herds along the water's edge or squeezing tightly onto ice floes. Although an individual mature walrus is more than a physical match for its ursine counterpart, a polar bear charge toward a gathering of walruses will almost invariably induce headlong flight into the water, a surging mass of elephantine blubber churning up a frothing foam of frigid ocean and providing openings of which an alert bear will seek to take advantage.

Ovsyanikov describes one instance in which a female on Wrangel Island "jumped down the small cliff and ran quickly along the front of the rookery, creating panic among the animals, all the while scanning the mass as if searching for something. Suddenly she rushed toward one group of fleeing walruses, mixed in with the rearguard for a moment, and snatched out a small calf."

The bear was just beginning to tear the skin off the calf's head when a large mature male bear appeared and, after brief resistance from the female, stole her bounty from her. So the female, clearly an adept hunter, repeated what she had done before, rushing the walruses until they stampeded for the sanctuary of the sea. "In the ensuing panic," wrote Ovsyanikov, "she made a short rush to the left and seized another small calf. It all happened so quickly that no walrus even tried to defend it."

There are no recorded examples of polar bears cooperating to hunt walruses—no reason, in fact, to believe that mature polar bears ever work together with intent at all. Ovsyanikov, however, has suggested that the attempts of bears to hunt adult walruses and the fact that those attempts almost all initially end in failure ultimately add up to the greater ursine good. "It is quite possible," he writes, "that a walrus that is repeatedly attacked and wounded is well on its way to an early demise, thus increasing the polar bears' chances of getting a meal."

When such a meal does come, Ovsyanikov suggests that the shared eating behavior that he observed on Wrangel serves a practical function, that of "opening up" the tough carcass and making it easier for all participating bears, including the one that killed the walrus in the first place, to gain their fill. But the social element of polar bear lives extends beyond the purely pragmatic.

While there is no recognized social structure among polar bears, there is an identifiable and noticeable pecking order, with mature males inevitably at the top, and subadults at the bottom, below even cubs with their mothers. At feeding sites, the arrival of a large male produces a noticeable stir among other bears, while subadults approach warily and cubs feeding alongside their mothers are sufficiently emboldened to make brief charges to keep them away.

And despite their largely solitary nature, individual bears will for a while pair up with others from their particular rung of the social ladder. They may rest together, sleeping close to each other on day beds they have fashioned for themselves out of snow as they wait for the sea ice to re-form. Polar bear mothers with cubs may spend time in each other's company. Subadult and even mature males walk together and play together, engaging in mock battles with each other, ritualized play fighting that may be quite rough and may last for forty-five minutes at a stretch, but in which harm is rarely done and injuries are almost never inflicted. Because of their awareness of their own power and that of their peers, polar bears take care to avoid full-blooded conflict with each other whenever possible, with one exception: when the time comes to mate.

• • •

The two bears have survived their subadult years and grown into fully mature males. In the process they have gone their separate ways and may never see each other again. But now, at five years old, they have more immediate and pressing concerns.

It is March; winter is drawing to a close, spring is around the corner, and the sea ice will soon begin to melt. In the snow that covers the floes, bear tracks reveal an extra purposefulness. The bears that have made them have not paused to sniff the air or meandered toward cracks in the ice; they have stridden forward, plodding relentlessly for sixty miles or more. These are the tracks of fully grown males who, across miles upon miles of flat, frozen wilderness, have caught the scent of a potential mate, apparently secreted through glands in the sole of a female's paw, and are driven to find her. When they do, they find other males in her presence, each with the same goal.

And so the battle begins.

Mating battles between mature males are rarely observed, but their ferocity can be inferred. For one thing, male polar bears are significantly more massive than females, an example of a phenomenon known to biologists as sexual dimorphism, in which males and females are dramatically different in size, appearance, or coloration. Almost invariably, when males are substantially larger than females, they are also significantly more numerous and must therefore compete with each other for the attention of a possible mate. There are reckoned to be, if anything, slightly more female polar bears than males, making the species the exception to the rule; but the long period over which females raise cubs means that during any given mating period, only approximately one-third of the mature females are available to breed. Competition, therefore, is intense, and further evidence of the severity of the fights that take place can be found in older males, which have far more scars and broken teeth than do females of the same age. Geoff York has told me that "I have seen two males, absolutely battered, exhausted, and emaciated, following a female to breed. The male in the rear of the group had one ear hanging to the side of his head and numerous slash and puncture wounds. A massive slash to the groin was large and deep enough to insert your full hand, yet this bear continued its pursuit." As he wryly observes, such behavior is why it is assumed males live shorter lives than females.

When one male has seen off his rivals and emerged victorious, he escorts the female away from the more heavily trafficked areas of prime seal habitat where the greatest number of bears is likely to be found, nudging and leading her to more secluded or remote spots, either high on a mountainside or far out on the pack ice. And then they begin to mate.

The couple spends several days together, during which they mate repeatedly. They copulate several times before the female releases an egg, and several more after that before the egg is successfully fertilized, a strategy that ensures the egg is not released without a mate being available and that it is not wasted on a mate that is unsuitable.

The female may mate with more than one male, and when she gives birth, each of her two or three cubs may be sired by different fathers. Whether she breeds with one or more, the end result is always the same. The two go their separate ways and likely will not see each other again.

The male wanders off, perhaps in search of another mate. In time the female will begin probing snowdrifts, looking for a suitable spot to make a den and give birth. But it will be a few months before she is ready to take that step. For now, she must eat.

Encounters

In the distance, near the horizon, there is movement. From where he stands, on the edge of the fast ice, the bear cannot quite make out the nature of the creature that is moving, or whether it has seen him. He raises his nose to the air, catches the scents that blow on the breeze: faint, and faintly familiar.

It is not a smell to which he is altogether accustomed, not the easily identifiable odor of a walrus on an ice floe or a ringed seal lying in its lair. But it is not entirely new to him. He has smelled it before, sometimes stronger, sometimes intermingled with a variety of fantastic and otherworldly sensory delights; in times past, on his way to the shores of the bay, waiting for the ice to freeze, he had dared to explore their provenance. He had been younger then, and hungrier, braver perhaps, more desperate certainly.

He had ventured forth in search of food, had been able to procure some scraps here and there, had even managed to avoid being seen in the process. Even so, as he became older and wiser, more accomplished at finding food where he was supposed to, where his mother had taught him and his brother so long ago, the need and temptation to take such risks disappeared.

He placed his front paws on a hummock to gain better elevation, but the scent was already starting to diminish. The creature was moving away. For a moment, the bear's predatory instincts urged him to follow, to stay upwind and to close the distance until he could make a decision on whether to attack. But it was an urge he could easily enough resist. The ice was fractured here, there were many leads, and seals were plentiful. There was no need to invest precious time and energy in stalking that potential prey. Besides, the effort might prove not only fruitless but potentially dangerous. Even on those occasions in the past when he had managed to snatch food scraps from the midst of such creatures, he had felt an uneasiness, a sense of danger always lurking.

Far better to turn back out to the pack ice. Let the creature take its scent off the fast ice and back onto land, unmolested.

That ours became the dominant species on Earth owes much to a series of circumstances, some confluent, some leading to others, some the product of good fortune, others of evolutionary adaptation: intelligence, a social nature, bipedal motion, an omnivorous diet, opposable thumbs, the use of fire and tools, an ability to adapt to a variety of climates and survive catastrophic events that limited or eliminated competitors and predators—ultimately, of course, our ancestors' development of agriculture and the knowledge to alter the environment to suit our tastes rather than being obliged to adapt to its whims.

Success was earned and prey was killed through teamwork and guile rather than brute strength; through all but the most recent of times, the risk of falling victim to those we could outsmart but not outmuscle remained all too real. When at night our children peer uncertainly over the bedclothes and complain of monsters in the corner or in the closet, it is likely a manifestation of a deep-rooted fear, an evolutionarily wise awareness that, lurking out of sight, predators larger and more powerful lie in wait.

Over time, as our ancestors grew stronger and more confident and as their ability to defend themselves improved, their attitudes toward those predators evolved, broadening from abject fear to encompass respect and even reverence. Our fellow species became not just potential predators but rivals, superiors yet also equals, and ultimately the subjects of art, tales, and rituals in whose drawing, telling, and enactment people sought to celebrate those creatures' spirit, to eulogize their power but also to demystify it, make it somehow less frightening.

For few animals was this truer than for bears. Their mystery and majesty, and the fear they engendered, were countered by their easily anthropomorphized features and their occasionally bipedal gait. In the centuries before primates were known to anyone north of the equator, it was common for human observers to imagine they were looking on our closest relatives, our spiritual kin.

There is a remarkable commonality in bear myths and tales, from Europe to North America and beyond. There are stories of bears becoming humans, humans becoming bears, bears raising human children, bears being raised as human children, women marrying bears and giving birth to their hybrid offspring. In some cultures, bears' imagined oneness with humans made eating bear meat as taboo as eating one's own kin; in others, doing so granted strength, but only if done with the appropriate degree of respect.

The Ojibway people of the Great Lakes region believed bears to be descended from people; the Cherokee held that bears had come into being when some of their ancestors chose to abandon the life of humans. As late as the eleventh century, Earl Siward of Northumbria and King Svend Estridsen of Denmark considered themselves to be descended from bears; in Siward's case, the veracity of his contention was said to be evident in the ursine appearance of his ears. A legend from the Ket people of Siberia held that bears came into being when a man disrobed to climb a tree, whereupon an unknown thief stole his clothes and the tree climber grew a bear pelt; although he looked like a bear, he could still comprehend human language. The Tlingit of British Columbia and southwest Alaska believed that not only could bears understand what people said, but their sensitive ears allowed them to hear from a great distance; accordingly, even a few carelessly uttered discourteous murmurs would be enough to prompt a bear to plot revenge.

Hundreds of miles to the north, the bear mythology of Eskimo and Inuit is characterized by a similar caution, an injunction against the demonstration of disrespect. To many Arctic coastal cultures, a polar bear cannot be killed involuntarily; it can only allow itself to be taken in order to enable its spirit to pass on to another dimension. The soul of a polar bear was said to linger for several days on the tip of the spear or harpoon that smote its physical form, observing the post-hunt rituals and dances, a knowledge that compelled the successful hunters not to mock, gloat over, or otherwise disparage the bear that had surrendered itself.

In his book *Lords of the Arctic*, Richard C. Davids writes that McGill University anthropologist George Wenzel, accompanying Inuit on a polar bear hunt in Resolute Bay in 1979, was told repeatedly not, under any circumstances, to ridicule or belittle a bear; doing so, he was assured, would bring bad luck. "Bears, the hunters kept telling him, were fully as smart as humans," reported Davids. (Indeed, Davids continues, some legends have it that, in the same way men recount successful polar bear hunts, the bears, too, regale each other with stories of stalking and killing those with two legs, whom they refer to as "the ones who stagger.")

It would seem an easy enough piece of advice to which to adhere, the admonition not to mock a polar bear; and yet it was one that Wenzel managed to transgress. As he watched his companions skin a bear that they had killed, he commented that polar bears were foolish in the way they allowed snowmobiles to approach so close before they began to flee. Two of those skinning the bear immediately stopped work, looked at him, and cautioned him not to speak of the bear in that way again. Two days later, the village chief quietly revealed that word had reached him of what Wenzel had said. He had, he reported, expected better of his visitor; he could only hope, he continued, that because he was a white man and thus assumed to be ignorant of the ways of the north, there would be no serious repercussions.

The imperative of respect, of not speaking ill of a polar bear either directly or behind its back, that informs the behavior of those on a hunt also permeates the tales told in a warm dwelling when the hunt is over. Like those recounted by cultures about ursids far to the south, Arctic stories of polar bears bestow human qualities on their protagonists and assume that humans and bears move effortlessly (if not always comfortably) between each other's worlds. It is an assumption that is a testament to the proximity of the tales' tellers to the natural world about which they speak, a proximity that is as true now for coastal peoples of the Arctic as it was several centuries ago for Europeans who recounted myths of the wild animals that lurked in the shadows of the forests.

Consider, by way of example, the following:

A polar bear fell in love with a married woman, and they began an affair. He warned her not to speak of their relationship to her husband, for fear that the man would surely seek to kill him. His concern was misplaced; the husband was a particularly poor hunter of bears, so much so, in fact, that, taking pity on her spouse's lack of success, the woman revealed her lover's whereabouts. The words, though whispered, traveled through the air to the ears of the bear, which left his dwelling before the husband arrived to seek vengeance for the betrayal. Reaching the woman's snow house, he raised himself on his hind legs, poised to smash through the roof with his front feet. But, at the last moment, he paused, dropped his legs to the side, and, sadly, wandered off into the distance.

Barry Lopez, who recites the tale in *Arctic Dreams*, notes the subtleties therein that speak to an intimacy and familiarity with polar bears and the Arctic, and a recognition of the danger posed by both. While to a European the notion of a mournful bear setting off on a long trek alone may seem, as he puts it, poignant, to an Inuk the ending is evocative on an entirely different level, summoning as it does the danger of being distracted in an environment where constant attention is a requirement and anything less poses perils.

The manner in which the bear in the tale raised up to smash through the roof of the woman's house is of course suggestive of the method polar bears employ to break into seal lairs, a small detail casually mentioned that would surely have escaped inclusion had the author or storyteller been anyone unfamiliar with the daily workings of *Ursus maritimus*. Such familiarity is a testament not to the supposed "oneness with Nature" too easily and frequently ascribed to anyone living in a nonindustrial culture, but, more prosaically, to that most valued commodity of any field researcher: the time and opportunity to make multiple, repeated observations.

Polar bears are in many respects like whales, not solely in that they are officially classified as marine mammals, or that they have been totems for environmental causes, but because they are both, at heart, wanderers. Whales and polar bears alike traverse great distances, through environments to which virtually all would-be human scrutinizers are singularly ill adapted.

It is a trial particularly vexing for anyone wishing to study cetaceans; for most who wish to observe polar bears, the challenge is not substantially less. But for those for whom, like the polar bear, the ice is a place of familiarity and plenty, there are fewer difficulties. Even then, however, the more frequent encounter is not with the bears themselves but with the signs of their recent presence, of their hunting or their passage, signs that, with the benefit of experience and observation, the astute can interpret with a greater degree of detail than the unaware.

"If you track a bear, you can tell whether it's a male or female by the size and dent of the [paw] imprint," says Bob Konana, an Inuk from Gjoa Haven in Nunavut, in a volume entitled *Inuit Qaujimaningit Nanurnut: Inuit Knowledge of Polar Bears*. "The female . . . has smaller feet and paws that are broader than the male's . . . and lighter in weight, while the male bear is much larger and weighs heavier than the females, their imprint tends to be deeper and wider."

It is also, says Konana, possible to tell from tracks whether the bear that has left them is healthy or unwell.

"The heel will dig more into the snow if it is skinny," he says. "If it is fat you can't see the heel bone."

Experience has taught Konana the most likely places for such tracks to lead.

"If you are following a pressure ridge and you see only old tracks," he says, "keep following it because newer tracks will always cross it at some point."

Pressure ridges — those areas where floes grind together, buckling under and rising over each other, creating miniature mountain ranges — are prime polar bear habitat. In their lee form snowdrifts; in the snowdrifts, ringed seals carve out lairs, and it is the smell of pups in the lairs that attracts the bears. But, says Konana, pressure ridges are not the only likely spots for polar bears; equally desirable are what the people of Gjoa Haven refer to in English as icebergs. They are not, however, icebergs as we understand them — chunks of freshwater ice that have calved from glaciers. In their native Inuktitut, Konana and his relatives dub them *piqalujat* — large, elevated expanses of old, hardened, thickened sea ice — and they say that the movement of these hardened floes, grinding through the fresher seasonal ice, keeps leads open and creates more openings for polar bears to hunt seals.

The reason why polar bears seek out pressure ridges or areas of broken ice is, of course, to feed. Frequent observation of these areas, noted and passed down orally over the course of centuries, has enabled Inuit and Eskimo to refine their own seal-hunting techniques—although some of the techniques that Natives insist they have seen bears employ in pursuit of their prey have not been widely reported in scientific journals. The notion that bears push blocks of ice ahead of them while stalking, or that they use such blocks—or even boulders—as tools to break through particularly recalcitrant breathing holes or even to pound unsuspecting walruses, engenders skepticism among Western researchers, even if oft repeated by Inuit and Eskimo observers. Konana asserts that he has seen bears use their intelligence in other ways.

"I have seen a polar bear [prepare] the seal hole," he says. "He cleared all the snow . . . around it, and made it really thin right to the ice cone over the hole. It was really thin, so that he could easily grab the seal when it came out."

And yet even inventive polar bears can meet with unexpected tragedy. The polar bear's head and neck are narrower than those of a brown bear in part to facilitate reaching into a seal hole and pulling a victim out onto the ice; a brown bear, were it to attempt the same trick, would almost certainly find its head stuck, but several of the residents of Gjoa Haven insist that polar bears are not immune to such setbacks themselves.

"Before there were so many snow machines, someone with a dogteam discovered a polar bear stuck in a seal hole," adds Konana. "It drowned. It was bloated with the gases of decay. It was stinky and . . . so full of gas the water was way down the . . . hole [from the pressure]. The bear attacked the seal and got stuck."

In a February 2009 edition of the science journal *Nature*, writer Richard Monastersky detailed a growing interest on the part of scientists in incorporating the traditional knowledge of polar peoples into their understanding of Arctic ecosystems and the change those ecosystems are undergoing. He quoted, by way of example, Elizabeth Peacock, polar bear biologist for the Canadian territory of Nunavut, as saying she often relies on Inuit friends for their insights into polar bear behavior.

"Recently, she was puzzled when a male bear with a satellite transmitter stopped moving for six weeks, acting like a female in her den rather than hunting seals as would normally be the case," Monastersky wrote. "An Inuit hunter solved the mystery by telling her that male bears sometimes rest if they are already fat and want to preserve their energy for the best seal season later in the winter. That's not an insight you'll find in the scientific literature."

For coastal peoples of the Arctic, modern accouterments have made life more convenient and secure than it once was. But the environment in which that life is lived remains hostile and inhospitable, and like the bears with which they share that environment, Inuit and Eskimo peoples must still adapt to its unique challenges. Even now, even in an age when flights may bring supplies of soda and microwave dinners, the beating heart that powers the culture of coastal villages in Arctic Alaska, Canada, Greenland, and Russia is subsistence, the consumption of the marine mammals that inhabit the sea ice and the water just beyond. In search of those marine mammals they, like their ursine compatriots, are wanderers, accustomed to traveling great distances for long periods. They, too, noting the ways in which polar bears wait for long spells by seal holes, adopt a means of hunting that is the personification of patience: lengthy periods of quiet anticipation punctuated with sudden bursts of frenetic activity. And while it is a life that is far from easy, in an environment that is far from forgiving, it is at least a milieu with which, after countless generations of habitation, coastal peoples of the Arctic are familiar.

In its seasonal rhythms and familiar natural cues, it is, if not exactly comfortable, at least comforting.

It is home.

For the Europeans who began to venture into what, to them, was a strange and terrifying world, it was something entirely different, about as far removed from the comforting embrace of home as it was possible to be. As Hugh Brody notes in *Living Arctic*, "Arctic and Subarctic societies must depend upon, and not simply exist in defiance of, the cold." Conversely, he continues, "Europeans, with their agricultural heritage, react with fascinated horror to the idea of a far northern cold." Or, as Jeannette Mirsky observed in her 1934 classic of northern exploration, *To the Arctic!*, "Only the very strongest of motives could induce men to undertake Arctic voyages during the period just following the discovery of the New World. Men were as fearful of the dangers of the Arctic as they were of the terrors of hell."

Those motives were, initially, territorial and financial, the two inextricably intertwined and manifested primarily in the search for a passage through the Arctic to the Orient, where it was assumed riches could surely be gained from the trade of teas, spices, gold, and silk.

However, far from providing an easily navigable passageway, the Arctic instead proved a formidable adversary, encounters with which provided more than ample justification for the apprehension it engendered in those who ventured into its depths.

A 1498 expedition led by John Cabot to the waters of Newfoundland simply disappeared, its fate never recorded or revealed. In 1553, two of the three ships of a much-heralded journey in search of the Northeast Passage failed to return; frozen into the ice off Russia's Kola Peninsula, the hulls inadequately protected against, and the crew improperly prepared for, the rigors of an Arctic winter, ships and sailors alike were found by fishermen the following summer. Every one of the complement had perished from, most likely, a combination of cold, starvation, and carbon monoxide poisoning as a result of burning coal in the wood stoves and sealing chimneys and portholes against the cold outside.

Forty-three years after that voyage, the Dutchmen Willem Barents and Jacob van Heemskerck led two vessels on a similar path to that taken in 1553; Barents had been beaten back by ice in the same region two years earlier but, notwithstanding the abbreviated ending of that endeavor, nonetheless remained confident of his ability to find and navigate the pathway to Cathay that he, and multiple others, assumed awaited discovery.

Sailing north from Norway, they became the first Europeans to spy Spitsbergen; but on this occasion, as during the previous attempt, sea ice proved an impediment to progress, sufficient to persuade one of the ships to turn south and head home. The remaining vessel—Heemskerck commanding, Barents the navigator, and Gerrit de Veer as mate—set forth anew, attempting to round Novaya Zemlya and head east, but the advancing pack ice of the Kara Sea encroached upon them, surrounding them and forcing them to retreat to a shallow bay they called Ice Haven. The name proved optimistic; even here, ice squeezed the hull of their vessel until the crew realized they had no option but to abandon ship and swiftly secure shelter on the shore. Using driftwood logs and planks from the crippled ship, they constructed a hut that they called the Safe House; there, all but one survived the winter, despite a near-death experience when they, like their seaborne predecessors almost five decades earlier, made the mistake of heating the cabin with sea coal and plugging up the chimney to trap the heat.

It was a frequently unsettling experience. The sailors were trapped miles from home, in an environment they did not know and in which they had not anticipated tarrying. They faced daily a conflict between the desire to keep their dwelling warm and the need to keep it aer-

ated. And they knew, too, that there were risks beyond the cold atten-
dant in opening the door, in the form of the foxes and, especially, the
polar bears that lurked expectantly outside, patrolling the darkness in
the near distance as the crewmen huddled together nervously inside.

On occasion, a particularly adventurous bear sought to break into
their haven. De Veer described the evening of April 6, 1597, when a
"beare came bouldly toward the house" and pushed forcefully against
the door:

> [B]ut our master held the dore fast to, and being in great haste and feare
> could not barre it with the peece of wood that we used thereunto; but
> the beare seeing that the dore was shut, she went backe againe, and
> within two hours after she came againe, and went round about and upon
> the top of the house, and made such a roaring that it was fearefull to
> heare, and at last got to the chimney, and made such worke there that
> we thought she should have broken it downe, and tore the saile that was
> made fast about it in many peeces with a great and fearfull noise; but for
> that it was night we made no resistance against her, because we could
> not see her. At last she went awaie and left us.

It was not their first close encounter. Two months previously, for
example, another bear advanced with apparently malicious intent
while several of the crew were outside, forcing them to retreat rapidly
toward their shelter and grab their firearms:

> We leavelled at her with our muskets, and as she came right before our
> dore we shot her into the breast clean through the heart, the bullet passing
> through her body and went out againe at her tayle . . . The beare feeling
> the blow, lept backwards, and ran twenty or thirty foote from the house,
> and there lay downe, wherewith we lept all out of the house and ran to
> her, and found her still alive; and when she saw us she reard up her head,
> as if she would gladly have done us some mischefe; but we trusted her

not, for that we had tried her strength sufficiently before, and therefore we
shot her twice in the body againe, and therewith she dyed. Then we ript
up her belly, and taking out her guts, drew her home to the house, where
we flead her and tooke at least one hundred pound of fat out of her belly,
which we [melted] and burnt in our lamps. This grease did us great service, for by that meanes we still kept a lampe burning all night long, which
before we could not doe for want of grease; and every man had means to
burne a lamp in his caban for such necessaries as he had to doe.

At times the crew must have wondered if polar bears would ever
leave them alone. On each of the last three days of May 1597, as the
shipwrecked men took advantage of the returning sun to work outside,
approaching polar bears forced them to retreat to the safety of their
house and the security of their weapons; each time, they shot the bears
dead. On the final occasion, however, the bear took its revenge from
beyond the grave:

Her death did us more hurt than her life, for after we ript her belly we drest
her liver and eate it, which in the tatse liked us well, but it made us all sicke,
especially three that were exceeding sicke, and we verily thought that we
should have lost them, for all their skins came of from the foote to the head,
but they recovered againe, for the which we gave God heartie thanks.

What de Veer and his fellow mariners could not know was that polar
bear livers are exceptionally high in vitamin A—a valuable vitamin,
to be sure, when consumed via a diet rich with carrots, broccoli, and
leafy vegetables, but potentially lethal when ingested at the kind of
levels found in the livers of polar bears (a consequence of the bear's
diet revolving around seals, which themselves carry what, to humans,
would be toxic levels of vitamin A).

Within two weeks of that episode, Heemskerck, Barents, de Veer,
and the others escaped not only the attentions of the ice bears, but

the confinement of Ice Haven. Their ship had not sunk but had been irredeemably damaged by the punishing ice, so the men piled supplies into two of the boats and gently threaded their way around the northern tip of Novaya Zemlya and into what is now the Barents Sea. A week later, that sea's eponymous discoverer died of scurvy; he and a crewmate who died the same day were taken ashore and hurriedly buried. The rest eventually — and remarkably — reached the coast of the Kola Peninsula and safety in the form of the ship that had begun the journey alongside them the previous summer.

De Veer's account of the voyage proved hugely popular and was widely translated; of his life following his return to land, little is known. It would not be especially fanciful to posit that he found contentment in the Arctic henceforth being a place to which he returned only in his memories, or that as he tossed and turned in his dreams at night, he more than once imagined himself confronted with a ravenous polar bear. For not only had he and his crew had to cope with their marauding attentions while shipwrecked; on Barents's previous voyage, on which de Veer had also sailed, an especially aggressive bear had both attacked some of his comrades and even set about eating them. This incident, too, de Veer recorded in detail in his book, beginning with the moment when, as two of the crew worked ashore,

A great leane* white beare came sodainly stealing out, and caught one of them fast by the necke, who not knowing what it was that tooke him by the necke, cried out and said, Who is that that pulles me so by the necke? Wherewith the other, that lay not farre from him, lifted up his head to see who it was, and perceiving it to be a monstrous beare, cryed and sayd, Oh mate it is a beare! and therewith presently rose up and ran away.

---

* In this adjective, perhaps, we can deduce an explanation for the bear's extraordinarily aggressive behavior.

The bear, at least in de Veer's telling, bit the man's head to pieces and began to suck out his blood as the rest of those ashore, approximately twenty in number, ran around in panic, some fleeing and a hardy few electing to try and save their comrade. They lowered their muskets and pikes and charged toward the bear, which, "perceiving them to come towards her, fiercely and cruelly ran at them, and gat another of them out from the companie, which she tare in peeces, wherewith all the rest ran away."

At this point a sizable complement of crew leaped into the boats and rowed furiously ashore, but not all of those already on the beach could be persuaded to attack the animal that had already slain two of their number. After all, they argued,

> Our men are already dead, and we shall get the beare well enough, though wee oppose not ourselves into so open danger; if wee might save our fellowes lives, then we would make haste; but now wee neede not make such speede, but take her at an advantage, with most securitie for our selves, for we have to doe with a cruell, fierce and ravenous beast.

Ultimately, three of the men took it upon themselves to attack the bear, but even after one of them, William Geysen, shot it in the head, the bear still did not let go of its quarry—although de Veer allowed that it staggered somewhat. Finally, he writes that Geysen struck the bear in the snout with his weapon, at which point it fell down, "making a great noyse," and Geysen, showing remarkable bravery given all that had gone on before, leaped upon it and slit its throat.

Few other explorers could recount tales quite so graphic or deadly. But for those who dared to venture forth into the Arctic's frigid embrace, as if it were not enough to have to contend with the slow physical and psychological torture of the cutting wind, cold air, and relentless attentions of ice floes, there was the knowledge that, immediately upon disembarking the sanctuary of the ship that had brought them to this unforgiving circle of hell, they were at risk of surprise attack from the predator whose terrifying nature was magnified by its silent approach.

"This fierce tyrant of the dills and snows of the north unites the strength of the lion with the untamable fierceness of the hyena," reads a description in an 1835 tome with the full and informative title of *The Mariner's Chronicle: Containing Narratives of the Most Remarkable Disasters at Sea Such as Shipwrecks, Storms, Fires and Famines, Also Naval Engagements, Piratical Adventures, Incidents of Discovery, and Other Extraordinary and Interesting Occurrences*. Continued the anonymous authors:

> This bear prowls continually for his prey; which consists chiefly of the smaller cecacia,* and of seals, which, unable to contend with him, shun their fate by keeping strict watch, and plunging into the depths of the waters. With the walrus he holds dreadful and doubtful encounters; and that powerful animal, with his enormous tusks, frequently beats him off with great damage. The whale he dares not attack, but watches anxiously for the huge carcass in a dead state, which affords him a prolonged and delicious feast: he scents it at the distance of miles. All these sources of supply being precarious, he is sometimes left for weeks without food, and the fury of his hunger then becomes tremendous. At such periods, man, viewed by him always as his prey, is attacked with peculiar fierceness.

"The annals of the north are filled with accounts of the most perilous and fatal conflicts of the polar bear," the narrative continues, citing by way of example an account documented by William Scoresby, whaling captain of great repute and renowned chronicler of the Arctic.

In his classic 1820 tome, *An Account of the Arctic Regions with a History and Description of the Northern Whale Fishery*, Scoresby relates the tale

---

* Cetaceans—that is, whales, dolphins, and porpoises.

of one Captain Cook of the *Archangel*, who during a 1788 voyage went ashore at Spitsbergen, accompanied by his surgeon and mate. Suddenly, a polar bear appeared as if from nowhere, leaping at Cook and seizing him between its paws. Doubtless seeing his life pass before his eyes, Cook yelled for the surgeon to fire his gun, which he did, shooting the bear in the head and killing it instantly.

Scoresby also tells a possibly apocryphal tale of an unnamed sailor who avoided death while being pursued across the sea ice, despite being unarmed. According to Scoresby, the fleeing mariner pulled off consecutively his hat, jacket, and kerchief, throwing each of them onto the ice behind him as he ran. The polar bear's curiosity apparently won out over its predatory instincts, causing it to stop and investigate each one as its would-be prey kept running until reaching the sanctuary of his ship.

The great polar explorer Roald Amundsen had already become the first to unlock the secrets of the Northwest Passage in 1906 and, five years later, the first man to reach the South Pole, when, in 1918, he set out in a ship of his own construction, the *Maud*. The plan was to sail west to east through the Northeast Passage, become frozen in the winter ice pack, and thence be carried with the drift toward the North Pole; but, having moored in the lee of an island north of Russia for the winter, Amundsen was bundled over by one of his sled dogs while descending the icy gangway from ship to shore that his crew had constructed out of snow. He fell heavily and broke his shoulder, an injury he would aggravate a month later when he was attacked by one of the island's residents.

This incident was triggered, as was the original fracture, by the exuberance of one of his dogs, in this instance a husky named Jacob which, as Amundsen stood on the shore with his arm in a sling, came screaming past him, yapping and whining, and tore up the gangway. In the darkness, Amundsen heard a heavy panting; out of the gloom, in slow but determined pursuit of the retreating canine, the panting resolved into an angry polar bear accompanied by a whimpering cub. The cub had presumably been teased or attacked by the dog, and now its mother apparently wanted revenge.

When the bear saw the human, she stopped in her tracks. Biped and quadruped sized each other up for several eternal moments. Amundsen decided to run toward the gangway, and the bear ran, too; it was, as Amundsen later described it, "a race between a healthy, furious bear and an invalid."

The invalid reached the bottom of the gangway, only to feel the bear's hot breath on his back and be flung to the ground by a powerful paw striking his bandaged shoulder. As the bear moved in for what Amundsen felt sure would be the killing blow, Jacob came to the rescue, racing back down the gangway and toward the cub. The mother wheeled around and set off after the dog, while Amundsen scrambled rapidly onto the ship. The following day, Jacob returned, frightened but otherwise no worse for the ordeal.

(Reflecting on the event sometime later, Amundsen dwelled not so much on the closeness of his escape but on the unexpected thoughts that passed through his mind in what he expected to be the final seconds of his life: "I had always heard that when a man faced the seeming certainty of death—as for a moment I did, lying at the feet of the bear—he usually had the mental experience of having all the chief incidents of his life pass before him in vivid and instant review. Nothing so serious or important occupied my thoughts as I lay expecting the death blow. On the contrary, a scene passed before my eyes which, though vivid enough, was certainly frivolous. I lay there wondering how many hairpins were swept up on the sidewalks of Regent Street in London on a Monday morning! The significance of this foolish thought at one of the most serious moments of my life I shall have to leave to a psychologist . . .")

 • • •

A century later, the prospect of an encounter with a polar bear is no less of an occupational hazard than ever it was for those who live in, work in, or visit the species' Arctic domain. It is why towns like Deadhorse and Barrow on Alaska's North Slope constantly advise their residents to be cautious when stopping outside and go to great efforts—through, for example, careful disposal of food waste—to prevent bears from being enticed into the community.

The advisability of such precautions was underlined in 1993, when a polar bear attacked and severely injured Donald Chaffin, a worker at a U.S. Air Force Defense Early Warning (DEW Line) site at Oliktok Point in Alaska, approximately thirty miles west of the oil installations of Prudhoe Bay. As Chaffin and a colleague watched television in the facility's living quarters, the bear appeared at the window, pressed its nose and paws against the glass, and peered inside. Chaffin, demonstrating the phlegmatic approach to wildlife typical of Alaskans, swatted on the window with a rolled-up magazine to shoo it away. Duly shooed, the bear showed up again shortly afterward, only to once more be deterred by the threat of Chaffin's periodical. On the third occasion, however, it would not be denied.

With one swift move, the bear crashed through the window; Chaffin and his colleagues leaped out of their chairs to flee, but as they fumbled with the door, the bear grabbed Chaffin from behind, throwing him to the ground and biting into his head and body. Two of Chaffin's colleagues grabbed fire extinguishers and trained them on the assailant; its focus now diverted, it dropped Chaffin and headed out into the corridor. As it struggled to gain purchase on the slick floor, the bear was shot dead by another of the workers, Alexander Polakoff.

Chaffin was rushed by helicopter to the hospital and admitted to intensive care; he survived but was permanently disfigured. He later sued the government for failing to provide safe housing, for prohibiting firearms from being kept on site (Polakoff had had to retrieve his shotgun from a hiding place in his bedroom because it was contrary to policy), and for allowing Natives to store whale meat just a few hundred yards from the living quarters. In the days leading up to Chaffin's mauling, polar bears had been spotted feeding off the meat; one of them, which had been prowling the area and peering into windows, had been shot dead just four days before the attack.

What happened to Chaffin was highly unusual. The victim was aware of the bear's proximity and the attack itself was dramatic and loud. More often, polar bears appear silently and suddenly, underlining the axiom told by Inuk guide Jimmy Memorana to researcher Ian Stirling: "If the bear is hunting, you won't see him until he comes for you."

In December 1990, Carl Stalker and his wife, Rhoda Long, eight months pregnant, left a family member's house in the northwest Alaska village of Point Lay late one night and set out on the short walk home. They rounded the corner and came face-to-face with a bear.

Long screamed. Stalker told her to run, pulled out his pocketknife, and ran the other way to lead the bear away from her.

Villagers found the bear two hours later, feasting on what remained of Stalker's carcass, and shot it dead.

In 2005, Eric Larsen and Lonnie Dupre were attempting the first-ever summer crossing of the Arctic Ocean sea ice to the North Pole, using a combination of skis and modified kayaks; they were doing so as part of a Greenpeace effort to draw attention to the impact of global warming on sea ice and hence polar bears, but the compassionate nature of their mission earned them no favors from one particularly persistent bear. Ultimately, the explorers were able to drive away the intruder with flares, but until they did, they found themselves on more than one occasion experiencing a closer encounter with the Arctic's iconic species than they had intended.

"I honestly think they were more curious in how we smelled and maybe whether they could eat us, rather than just full-on hunting us," reckons Larsen in hindsight. "But they were definitely being careful in how they approached us. That first bear we saw, that was one of the scariest experiences of my life, because it was coming straight toward me, on its belly, pushing snow in front with its nose. I had my hood up, I was working on my boat, we were both bent over, arranging gear in our boats, and Lonnie goes, 'There's a bear.' It was on my side of the boat, coming straight toward me. I was freaked. You know, the wind was blowing, it could have gotten within a foot of me quite easily. Or even closer. I was oblivious. I wasn't looking around, and even if I had just casually glanced, I probably wouldn't have seen it. It was so well blended in."

That same night, another bear—or, just as likely, the same one—returned while Dupre and Larsen were cooking in their tent.

"When the stove is going in that tent, you can't really hear much," Larsen explains. "We were chatting away and laughing, cooking, and then literally Lonnie turned the stove off and we heard a step. Then we heard the boat move; it pushed the boat into the tent."

Dupre had placed his camera bag in the boat; the two men listened as the bear picked up the bag and dropped it.

For a moment, there was silence.

Then the bear jumped on the tent.

The men leaped to their feet as best they were able; their sudden movement and their yells caused the bear to back off, and the flares did the rest. Which goes to show: polar bears can be deterred.

Since 1990, Nikita Ovsyanikov has been studying the polar bears of Wrangel Island in the Russian High Arctic; there can be few people in history who have experienced as many close encounters with polar bears, and while most of those encounters have been placid, the sheer amount of time he has spent in the animals' proximity all but dictates that at least some have been tense affairs. Even so, the researcher, who refuses to carry a firearm, has managed to preempt possible attacks with more basic weapons.

On one occasion, Ovsyanikov grew increasingly uneasy as a huge old male advanced toward him. "He showed no signs of aggression, no fear, no excitement, no emotions at all. He behaved like a giant who has noticed an insect in his path and has decided to take a closer look." When his tried and tested methods of deterrence — such as waving a shovel high above his head — failed to show any sign of halting the bear's advance, Ovsyanikov threw a rock. The missile landed on the hip of the bear, which stopped, appeared confused, and looked behind him before taking a few more steps forward. Ovsyanikov threw another rock, which struck the bear's shoulder.

"And then this giant once again behaved in a manner I had completely failed to anticipate," the scientist wrote. "He suddenly got terribly frightened."

The bear fled for the safety of the water, so fixated on his goal and on retreating from the unseen force that had hit him that he ran through the middle of a wall of walruses.

"One time, I was behind about a three-inch-thick steel door, and a big bear came and made a beeline for where I was, knew exactly where he was going, didn't hesitate, and put both his front paws on that door and proceeded to march in," recalls Robert Buchanan, who as president of the not-for-profit Polar Bears International has spent many years among the polar bears of Hudson Bay. "I had no idea he was coming through, of course, and the only thing I had at my disposal was a broomstick, and I just started beating the living piss out of his nose just as hard as I could and as fast as I could, and he looked at me like he thought I was stupid. And I just kept whacking it, and he started to back off, like, 'Well, this is not worth the trouble,' and crawled back out."

By and large, reckons Buchanan, polar bears generally aren't that interested in humans as potential victims. Unlike blubber-coated seals, he says, "there's just not that much on a human being to interest them." But on those occasions when a polar bear does attack, it is often as not exceptionally hungry—as in the case of the bear that killed Stalker, which, according to the vet who examined it, had "no more than a few ounces of fat" on its body—or very young (or perhaps both, as the latter condition is often an indicator of the former).

"The ones who give you the most trouble, just as in our world, are teenagers," Buchanan continues. "Teenagers in the polar bear world usually don't have enough nourishment or are not smart enough yet to understand the risk for them, but they are going to be the meanest and most aggressive. One thing adult polar bears won't do, and subadults haven't always learned this yet, is they will not attack unless they're pretty sure they're going to get that person, or that seal. They will not waste that energy unless they're absolutely comfortable they're going to make that score. So when they put their ears down, you're in deep doo-doo."

University of Calgary professor Stephen Herrero, author of the seminal volume *Bear Attacks: Their Causes and Avoidance*, agrees: "A typical encounter with a polar bear, it is not going to be interested in injuring you or hunting you," he says. "But, in the event that a polar bear does decide you look like something to eat, then you'd better be well prepared, because they're a difficult bear to deter once they get into predatory mode, and they can get into that mode so suddenly."

In a 1990 study, Herrero and Susan Fleck found that, of twenty documented fatal or injurious polar bear attacks in Canada between 1965 and 1985, fifteen were apparently the result of predatory behavior, three an apparent case of a mother defending her young, one apparently a combination of both, and one of unknown motivation. Of those predatory fifteen, thirteen were identified as being by males, of which seven were subadult and four were described as "thin" or "skinny." As was the case with the bear that attacked Chaffin, eight of the twenty involved attractants such as garbage, animal carcasses, or food, and another was near an Inuit hunting camp.

"Bears are curious by nature and make a living by exploring possible things to eat," Herrero points out. "And so managing odors is one of the biggest factors in improving your safety."

In general, Herrero and Fleck wrote in their paper, their data "support the conclusion that polar bears, especially males, can be predators on people. The data also show that such events are rare."

Others put it more strongly.

"People talk about polar bears stalking and hunting humans," said Tom Smith, a polar bear researcher at Brigham Young University, as we sat eating dinner one evening on the shores of Hudson Bay. Outside, a light dusting of snow was insufficient to hide the solitary bear that lay a couple of hundred yards away in the fading daylight. "Well, if that's so, they're doing a pretty piss-poor job of it."

In the previous 125 years, Smith pointed out, polar bears had killed 8 people in Canada and 2 in Alaska. In recorded history, they have killed 19 in Russia. Conversely, of the 353 polar bear–human interactions Herrero and Fleck catalogued in Canada, 14 resulted in human injury and 6 in death. But as a direct result of those encounters, "at least 251" of the polar bears were killed. A separate study counted 50 serious encounters between polar bears and people on Svalbard; 1 of the people involved died, and 46 of the bears.

When Europeans first penetrated the Arctic's icy depths, they did so often with a sense of adventure but also with foreboding—a foreboding that almost certainly only grew as awareness of the Arctic's hostile nature spread and intensified, as reports from returning vessels spoke of the horrors and dangers the region presented, and as other vessels failed to return altogether. In that context, a profound apprehension of the predator that lurked silently and threateningly in the shadows was entirely understandable, and a resolve on the part of sailors to defend themselves, even proactively, no less so. But gradually, the motivation morphed from fear to necessity and desire—for warm fur to protect them in a cold climate, for food to eat and fat to burn—and thence, ultimately, to fun; instead of shooting polar bears out of self-defense, Arctic explorers shot them because they could. "Europeans took to killing any bear they saw," writes Barry Lopez. "They shot them out of pettiness and a sense of rectitude. In time, killing polar bears became the sort of amusement people expected on an arctic journey."

As early as 1585, the men of an expedition commanded by John Davis in search of the Northwest Passage came upon "white bears of a monstrous bigness," whereupon they, "being desirous of fresh victual and the sport, began to assault them." The following day, they found a polar bear sleeping peacefully. So they shot it in the head, "whereupon we ran all upon him with boar spears and thrust him in the body."

Lopez recounts a tale told by William Scoresby, in which walrus hunters set fire to a mound of blubber on the ice specifically to attract polar bears. A female and two cubs approached, the mother leaving the cubs at a safe distance as she approached the fire and tried to hook flaming pieces of blubber from it. Seeing the difficulty she was experiencing, the crew threw her some small pieces, which she took back to the cubs. And then they took aim at first one cub, and then the other, and shot them both. For thirty minutes the mother "laid her paws first upon one, and then the other, and endeavored to raise them up." She called to them, touched them, looked upon them "with signs of inexpressible fondness."

So the sailors shot her, too.

Scoresby tells another tale of how, in 1812, another mother bear, also with two cubs, approached a ship he commanded. This time, the sailors shot the mother and captured the cubs, which, "though at first evidently very unhappy, became at length, in some measure, reconciled to their situation; and being tolerably tame, were allowed

Mature polar bears are mostly solitary.

Robert and Carolyn Buchanan / PolarBearsInternational.org

A polar bear mother and two cubs rest in the snow.
Robert and Carolyn Buchanan / PolarBearsInternational.org

Two young males spar in the snow near Canada's Hudson Bay. Although playful, such sparring can sometimes inflict bloody wounds.
Robert and Carolyn Buchanan / PolarBearsInternational.org

Cubs stick close by their mother's side for the first two years of their lives.
Robert and Carolyn Buchanan / PolarBearsInternational.org

A polar bear's forepaws are massive – ideally suited to act as snowshoes, paddles in the water, and powerful weapons with which to strike seals. *Kieran Mulvaney*

Polar bears are perfectly at home in the water for short periods.
Nick Cobbing

A ringed seal about to surface at a breathing hole in the ice. The outlines of a net used by researchers to capture seals for tagging and measuring can just be seen at the hole's perimeter. *Brendan P. Kelly*

In Churchill, Manitoba, the "Polar Bear Capital of the World," residents
and visitors alike are warned to keep their eyes open.

Robert and Carolyn Buchanan / PolarBearsInternational.org

The Tundra Buggy Lodge near the town of Churchill, on the shores of Canada's Hudson Bay. *Robert and Carolyn Buchanan / PolarBearsInternational.org*

Close encounter: A polar bear investigates the occupants of a Tundra Buggy near Churchill. *Dan Guravich / PolarBearsInternational.org*

A skinny polar bear photographed off the Alaska coast. As sea ice melts and retreats, scientists predict polar bears will find it harder to feed and survive.
Kieran Mulvaney

Polar bears feeding at the town dump in Churchill. Since the dump was closed, instances of so-called problem bears have decreased dramatically.
Dan Guravich / PolarBearsInternational.org

A polar bear appears to stare at its own reflection in the water off the coast of Greenland. *Nick Cobbing*

occasionally to go at large about the deck." That didn't prevent one of them from trying to escape, leaping overboard onto the ice, attempting to wrest from its neck the noose that tethered it to the deck and to run as far from the ship as it could until, realizing it was held fast, "he yielded himself to his hard necessity, and lay down on the ice in angry and sullen silence."

The fate of those bear cubs is not reported, but that of other polar bears captured alive is a matter of public record—and indeed has been for a surprisingly long period of time. Ptolemy II, king of Egypt from 285 to 246 BCE, purportedly kept one in his private zoo in Alexandria; Roman emperors acquired at least some and on occasion deployed them against gladiators; and the Japanese emperor Korehito received a pair as a gift upon his accession in 858 CE. The provenance of such bears is unknown; more recently, in 1928, the U.S. Coast Guard cutter *Marion* encountered—again—a female with two cubs, whereupon the crew shot the mother and one of the cubs, hauling them aboard for food and fur. They retrieved the terrified, angry cub that remained, stowing her first in the fore hold and then, after she contrived to escape, in a cage that was swiftly fashioned for her. Named after the vessel on which she was forcibly transported, Marion was ultimately delivered to the National Zoo in Washington, DC, where she lived for twenty years.

By then the great age of Arctic exploration had passed—as had, more significantly for the fate of polar bears, the commercial whaling industry that had both resulted from and inspired much of that exploration. Fewer men setting sail on lengthy expeditions into the deepest depths of the Arctic ice pack for months or years on end meant fewer men to come into contact with polar bears and shoot them on sight. But the world by then was also growing rapidly smaller, and the Arctic, like much of the rest of the globe, was increasingly within reach.

The 1940s saw airborne hunters flying to Alaska to shoot polar bears for sport and bring back their hides as trophies to hang on the wall or place on the floor. Over the course of just one year in the 1950s, the farthest distance from the Alaska shore that a polar bear was shot climbed from 60 miles to 200; during that decade and the next, airborne hunters reportedly on occasion strayed so close to the Siberian coast in their search for quarry that they prompted the scrambling of Soviet fighter jets. By the 1960s, the Norwegian Travel Service was declaring the polar bear hunt on Svalbard to be "one of the finest big game hunts in the world . . . hunting them in the pack ice is an experience a hunter can never forget," while a Soviet biologist was lamenting that there were now more polar bear hunters in the Arctic than there were polar bears for them to hunt.

In 1965, the *New York Times* was moved, in an editorial, to fulminate against the burgeoning Alaska hunt:

> The polar bear is a victim of a peculiar—and particularly repulsive—expression of man's egotism. Wealthy men have taken to hunting bears in Alaska from airplanes. Two planes are used to herd the bear to an ice floe suitable for landing. While one plane lands and the hunter gets out, the other plane maneuvers the bemused animal within the hunter's gun sights. More than 300 polar bears, a new record, were killed in this fashion last winter in Alaska.
>
> This kind of hunt is about as sporting as machine-gunning a cow.

The trophy hunt in Alaska was, however, the epitome of decorum compared with the situation in Norway. There, polar bears were being killed not just for sport but also because the sealing industry saw them as a threat to their livelihood; consequently, the methods used to bring about their death were ruthlessly efficient rather than adventurous. Set guns were placed in wooden boxes on small platforms; approaching bears would seize the bait that had been thoughtfully left for them, triggering the guns, which shot them at point-blank range in the face.

That same year, Alaska senator Edward "Bob" Bartlett had addressed his colleagues with his growing concerns:

> I am informed that there are no accurate or reliable figures on the total world polar bear population or the size of the annual kill. Scientists know very little about the habits of habitat, reproduction, longevity, or population structure. They do not even know the answer to the basic question of whether there is but one population of polar bears moving from nation to nation on the slowly revolving ice pack, or whether there are two or more populations. Are there Soviet bears and American bears, Danish bears and Canadian bears, or are there just the bears of the world?

In conjunction with then–Secretary of the Interior Stewart Udall, Bartlett called for ". . . an international conference of Arctic Nations to pool scientific knowledge on the polar bear and to develop recommendations for future courses of action to benefit this resource of the Arctic region."

That gathering, the First Scientific Meeting on the Polar Bear, convened in Fairbanks, Alaska, in September 1965, its forty-six expert participants in general agreement that continued indiscriminate shooting would likely, at some point, have an impact on polar bear numbers, if it was not already doing so. But how much impact?

There was, as Bartlett had noted, no agreement on whether there was but one circumpolar bear population—as advanced by Soviet representatives at the meeting—or several, the view generally held by others. Whether divided into one population or several, there was nothing resembling consensus on how many polar bears in total roamed the Arctic. Five thousand, said some. More than 10,000, reckoned Canada's Richard Harington. Between 10,000 and 19,000, according to official U.S. estimates. As high as 25,000, according to others. Nor was there an accurate accounting of the number that were being killed, although a quick calculation by those in the room produced a figure of 1,475 annually, by sealers, sport hunters, and Natives alike. Depending on which population estimate you favored, that was anything between 6 and 30 percent of the world's polar bears being shot every year. Even given the most optimistic assessment of their numbers, that was unlikely to be sustainable.

Acknowledging the possible existence of a problem was one thing. Agreeing on a way forward was another. The Soviet Union had banned hunting in 1956 and in Fairbanks urged the other Arctic states to follow suit; but there was much money to be made in the United States from the Alaska hunts, the Norwegian representatives insisted their nationals had to kill bears to protect the country's sealing industry, and Canada and Denmark (the latter of which administered Greenland) argued that any ban would violate the long-established rights of Inuit to engage in subsistence hunts. A fallback proposal, for a five-year hiatus during which denning areas could be mapped and sanctuaries established, was met with scarcely more enthusiasm.

Not that the Soviets themselves embraced all ideas placed on the table. When the United States floated a five-year international circumpolar research program, in which bears would be fitted with radio collars that could be tracked by satellite, the Soviet delegation—noting the presence of representatives from the Pentagon and State Department at the meeting, and noting further that the man giving voice to the notion was an officer of the United States Navy—raised the fur on its collective back, issued a feline hiss, and ensured that such a suggestion would not be re-aired for the duration of the conference.

At the meeting's close, participants were united in agreement that polar bears were an international circumpolar resource—a word that has fallen increasingly out of favor because of its perceived exploitative connotations but in this context was meant to demonstrate only that, alive as well as dead, the species was of value to, and merited protection by, the entire global community. They agreed also that cubs, and females accompanied by cubs, should be protected at all times. Beyond that, however, there was mostly despair that "scientific knowledge of the polar bear is far from being sufficient as a foundation for sound management policies."

Those assembled cautioned that each nation in the polar bear's range should "take such steps as each country considers necessary to conserve the polar bear adequately until more precise management, based on research findings, can be applied." And they urged that each of those countries should also, as a matter of priority, "conduct to the best of its ability a research program on the polar bear within its territory or adjacent international waters to obtain adequate scientific information for effective management of the species."

The Fairbanks meeting was a rock in a pond; the ripples that spread ever outward included a burst of cooperative field research, from Hudson Bay to the Beaufort Sea, on Svalbard and in East Greenland, and the establishment of a Polar Bear Specialist Group (PBSG) that provided a forum for that research to be discussed and dissected, its conclusions challenged and debated. The group's first meeting—held behind closed doors in anticipation of possible political overreaction to discussion of international cooperation concerning, or national jurisdiction over, a polar bear population that might have the temerity at the height of the Cold War to wander from the sea ice of the Eastern Bloc to that of the West—was held in the appropriately neutral venue of Morges, Switzerland, in 1968; two years later at the same location, the group convened again.

In 1970 as in 1965, the Soviet Union put forward a proposal for a five-year moratorium on all hunting; as in 1965, the proposal did not receive universal agreement, but unlike in Fairbanks, resistance was not absolute. In an illustration of the extent to which the ground had shifted over the previous five years, the group asked governments to examine their management programs "with a view to drastically curtailing the harvesting of polar bears beginning the next hunting season and extending for the next five years."

With the hunt in some areas threatening to escalate out of control, several Arctic nations heeded that call. Prior to 1971, the number of polar bears an individual could kill in Alaska was without limit; that year the United States imposed a limit of three. The number of sport-hunting permits was capped at 210 for the western area of the state and 90 for the north. The following year, the United States adopted the Marine Mammal Protection Act, and at a stroke, sport hunting of polar bears, which two years previously had been essentially a free-for-all, was banned.

In 1970, the Norwegians limited the number of sport-hunting permits in Svalbard to 300. They had already, in 1965, banned the shooting of females and cubs and, two years after that, the hunting of any polar bears from snow machines, aircraft, or boats. In 1973, Norway subjected all polar bear hunting on Svalbard to a five-year moratorium. To date, that moratorium has not been lifted.

Momentum was now building irrevocably behind an international treaty—a suggestion first made by the Soviet Union at the 1970 meeting. The Soviet motivation, it later transpired, was not so much anxiety over polar bears' well-being—from their perspective, the hunting ban they enacted in 1956 had essentially taken care of their responsibilities on the matter—but more of a desire that the five Arctic states be seen as being the ones with authority and jurisdiction over the north polar regions. The concern, in other words, was geopolitical rather than environmental, but the effect was the same. By the time the Polar Bear Specialist Group met for a third time, in 1972, early drafts of a treaty were already being circulated, and it became increasingly clear that the only doubt about its adoption was its timing.

A number of concerns were raised and addressed. It was important, it was agreed, that a new agreement contain nothing that contravened any wider, existing international treaty. The treaty would be open to signature and ratification only by the five Arctic states; accordingly, it would need a provision under which those five nations would be expected to pressure nonsignatory nations and nationals not to hunt polar bears on the sea ice that covered international waters. There would need to be exceptions: it should not, for example, be a violation of the treaty to kill a polar bear in self-defense. Most important, the agreement would need to include particular emphasis that it did not in any way affect the rights of indigenous peoples to conduct traditional subsistence hunts.

Even with those caveats, the International Agreement on the Conservation of Polar Bears and Their Habitat, as signed in Oslo, Norway, in November 1973, clocks in at under a thousand words, almost half of which are concerned with organizational matters—who can ratify, when, and where. The most immediately relevant of them all were the first eight in Article I, which spell out with impressive clarity and brevity the agreement's singular mandate. Notwithstanding the negotiated exceptions—science, self-defense, traditional rights—the article effectively brought to an end the era of commercial polar bear hunting.

"The taking of polar bears," it stated unequivocally, "shall be prohibited."

Perhaps just as significant in the long term, however, was Article II, which instructed the signatories to "take appropriate action to protect the ecosystems of which polar bears are a part, with special attention to habitat components such as denning and feeding sites and migration patterns."

This provision, too, was enacted with great rapidity. The same year the agreement was signed, the same year that Norway placed a moratorium on polar bear hunting in Svalbard, Oslo also protected approximately 40 percent of the archipelago's landmass by royal decree. In 1976, the United Nations Educational, Scientific and Cultural

Organization (UNESCO) designated the Northeast Svalbard Nature Reserve as a Biosphere Reserve, as a result of which most of the denning areas and important summer sanctuaries in the area are completely protected.

And so it continued. The Northeast Greenland National Park, 375,000 square miles in area, the largest national park in the world, was established in 1973, acknowledging traditional rights by allowing Inuit from neighboring settlements to hunt there, but protecting the objects of their attention by requiring that such excursions be confined to the distance a sled could travel in and out in the course of a day. In 1976, the Soviet Union designated Wrangel and Herald islands as State Reserves, off-limits to virtually all visitors.

The agreement, wrote Ian Stirling, represented the first time the five Arctic nations had worked together to negotiate an agreement on a circumpolar issue; and, he continued, there was "still no other polar subject upon which the circumpolar nations have come to mutual agreement."

Almost overnight, the nations in which polar bears live ceased to regard those bears solely as a menace to be shot on sight, or as targets to be aimed at for sport and profit. At a time marked by a growing global environmental ethic, they placed themselves at the forefront, both recognizing that polar bears' presence in their territories conferred upon them a responsibility and immediately choosing to exercise that responsibility.

The agreement's adoption and success are all the more remarkable for the fact that it is held together almost entirely by the will of the countries that signed it. There is no enforcement mechanism, no infrastructure to oversee compliance, nothing to compel adherence other than the collective will of the countries involved.

But even as direct pressures eased, as polar bears no longer needed to fear being pursued across ice floes by rich men looking to display their bravery in the form of a rug,* as mothers and cubs could curl up in the warm darkness of their dens safe from the attentions of bipedal predators, other threats emerged—invisible, insidious, and with a potential impact far greater even than a parade of aerial sharpshooters could ever hope to have.

It had been a long summer. Several months of heat and nothing to eat do not a contented polar bear make. He found, as in previous years, an earthen den, sunk deep into the permafrost by generations of polar bears before him; its shade lowered the temperature enough that it was not sweltering, and its shelter protected him from the insects that buzzed outside the entrance. But now there was a familiar feeling in the air, a crispness that heralded the return of the ice and of the seals that would replenish his depleted fat reserves. The snow was now thick on the ground, deep enough that he could roll in it, rub his snout in it, toss it over himself. Its coolness was refreshing and comforting. Its familiarity reminded him of the good days that its arrival had always heralded.

He had left the den behind him, was traveling along a path similar to the one he always used, following the trails and scents with which he was familiar. He was determined but his pace was measured; the ice would not freeze up any quicker if he hurried, and he had used up

---

* However, although sport hunting is vastly reduced, it has not been entirely eliminated, as Canadian Inuit are permitted to sell some of their subsistence quota to non-Native hunters. By 2006, the portion of the quota from sport hunting had increased from 1 to 15 percent since the 1970s, although it seems likely to decrease following restrictions placed by the U.S. government on the import of polar bear products into that country. There is some concern also over increases in some Inuit quotas and catches in Canada and Greenland, particularly given that the species is facing additional threats from, primarily, climate change. Additionally, although polar bear hunting has long been banned in Russia, illegal hunts are believed to be taking place on a scale of great concern to environmentalists and scientists.

energy reserves over the summer without being able to acquire any more. Far better to take his time, even to stop now and then to roll over onto his back, close his eyes, and feel the comforting blanket envelop him.

He lay there for a while, his massive paws dangling in the air, projecting an image more akin to an oversize pet than a ravenous predator. But then, with a start, he opened his eyes.

That noise.

He had heard it before, several years ago, and even now he could not be entirely sure what exactly had happened. There had been a pursuit of some kind, a fogginess, a disturbed sleep, the feeling of waking up yet not being able to move.

It had begun with that same noise. Yes, that noise was familiar, and it was growing louder.

He rolled back onto his feet and began to run—not a flat-out gallop, but a loping stride, enough to carry him into the willows and out of the open.

The object was right above him now, its clattering sound growing louder and louder. A slightest of stings in his hip briefly caught his attention, but he continued onward. Suddenly, though, his legs felt heavy, his paws started to act as if they were trying to obstruct each other. He felt an overwhelming urge to sleep. The willows, so close, seemed as if they would never get any closer. He tried to churn his legs forward some more, but the snow seemed to grip his ankles and pull him down.

Then everything went black.

"The bigger adult males don't really care about the helicopters," says Jon Talon, a helicopter pilot who carries tourists and scientists in search of the polar bears of Hudson Bay. With tourists on board, he remains at a discreet distance, far enough away that the bears don't even notice his proximity; scientists he brings down to the bears' level, low enough to fire a sedating dart into the animals' flanks.

"They all have personalities, they all react differently, even the big adult males," he continues. "I had one of them try to jump up and grab the helicopter. We have adult males that we just fly up to, they just sit there, we put the dart in them, they go to sleep, an hour later it just sits up, and when we fly over there again it's still in the same spot, just hanging out. Other bears, they run as soon as they hear the helicopter. Obviously, we take precautionary measures; if there's a bear running and it's obviously very terrified of the helicopter, we're not going to chase it down and dart it. These guys, the researchers, they have thirty-five years' experience in the field, some of them. They know these animals."

Which is not to say that only those bears that sit passively are the ones to be darted; to do so would result in a wealth of data from the biggest, boldest, bravest males and an absence of information about the rest. Even if not exactly terrified, many bears—often, Talon notes, the ones that have been tagged before and that therefore associate the descending helicopter with some form of unpleasantness—will jog away and glance anxiously over their shoulders. After all, notes polar bear researcher Geoff York, "it's the closest thing to an alien abduction that you or I could imagine."

Once a bear is spotted, its size is estimated and the appropriate amount of tranquilizer—Telazol: part Valium, part paralytic—calculated. If the bear does take flight, pilot and biologist seek to work it toward a spot that is relatively flat and as far as possible from water. Polar bears may be creatures of the ice, but it is to water that a frightened polar bear will instinctively flee, an outcome to be avoided at all costs when that bear is sedated.

The helicopter descends to about twelve feet—close enough that if a bear were so inclined, it could leap up and grab the skids—slides in behind its quarry, and angles slightly to the right, at which point the darter fires the shot and the aircraft moves off and waits for the drug to take effect. Depending on the size and condition of the bear, the drug can take as few as three or as many as twelve minutes to take effect. Once it does, the bear falls asleep, the helicopter lands, and the biologists get to work.

Time is short: although the bear will be to some extent incapacitated for as much as three hours, it will be immobile only for an hour or so, and it is during that period that the researchers must fulfill most of their tasks.

Priority one is to tag the bear, in four separate places: tattoos on the upper and lower lips and tags in the right and left ears. "Every bear gets four markings because you'd be surprised how many times we recapture a bear and can barely read its markings," York explains. "Polar bears live pretty tough lives; ear tags get chewed on by cubs, and in males they get damaged in fighting."

The bear is measured: its length, its girth, its paws, its weight. If it is the first time a bear has been captured and tagged, the researchers pull a tooth — always a vestigial premolar, a relic tooth that is no longer needed and many bear species no longer have. Back at the lab, the tooth will be cross-sectioned and dyed, unveiling a series of rings that, like those of a tree, reveal the animal's age. Blood and serum samples are examined for clues to the bear's health and diet and are kept in storage for tests that may in the future be deemed useful and have not yet been considered.

When measuring and bagging one bear, a team of two scientists — frequently enlisting the help of the pilot — takes about an hour to run through the checklist, watching all the time to be certain the subject isn't overheating or otherwise suffering. With multiple bears, such as a mother and cubs, the same process may take an hour and a half. When all is done, the researchers spray a number on the pelt: an advisory, until it disappears when the bear molts, that this individual has been captured and tagged already this season and should not be so again, and a caution flag to any Native hunters that it has been injected with a small amount of tranquilizer, which is often cause for them to allow it to pass by.

If the process seems intrusive, it is, for many polar bear biologists, the extent of their hands-on experience of the animals they study: a view from an aircraft, a tranquilizer dart, an hour or perhaps two measuring and weighing, and the opportunity, while on the ground or in the air, to infer behavior from clues — tracks here, a seal carcass there.

Facts and figures about how far and where polar bears roam are derived more from plotting GPS signals from radio collars than from direct observation. The inaccessibility and inhospitability of the polar bear environment, coupled with the sheer extent and unpredictability of an individual bear's movement, make prolonged observations in the field all but impossible.

"There are some places where you can do direct observations, but direct observations can only give you so much information, and they're only possible in certain very limited portions of the polar bear range," explains Steven Amstrup of USGS. "So, for example, by going out and staying out in a cabin where there are polar bears, and watching them, you can understand certain things about their behaviors and watching them. If you're lucky, like Ian [Stirling] was when he was watching polar bears feeding, he was able to watch them hunting seals; that's not possible most places. But even that only gives you a glimpse of certain aspects of the polar bear lifestyle. It doesn't tell you anything about population dynamics, for example.

"Almost all of what we know about polar bears we know from capture/recapture: flying out in helicopters, darting them, catching them, ear-tagging them, and then doing it again."

But, while the nature of polar bears may seem to preclude the emergence of an Arctic ursine Dian Fossey, there are a few locations where they congregate at the beginning and end of each season, locations where, for at least certain periods of the year, a dedicated biologist can base himself or herself, where he or she can watch and learn and develop an awareness of polar bear behavior that cannot be ascertained from any radio collar or tooth cross section.

On Wrangel Island, Nikita Ovsyanikov continues to this day the long-term study he began in 1990. And on the southern shore of Hudson Bay, for over forty years researchers have watched at Cape Churchill as polar bears gather en masse before heading out onto the newly frozen sea ice.

Wrangel Island is forbidding and remote; it took several days of grinding through thick sea ice for the *Arctic Sunrise* to reach its coast in 1998. Ten years later, a trip to Hudson Bay required nothing more onerous than a ticket to board a train.

That accessibility makes southern Hudson Bay and its environs unique. It is more than a location from which scientists can watch polar bears up close. It is the only place on Earth where, for several months a year, the worlds of polar bears and humans intersect, where tourists join locals for a glimpse of an animal they could not hope to see anywhere else. It is where a young polar bear is shaking off the effects of a tranquilizer as it heads toward the coast while, several hundred miles to the south, a train travels slowly through the night, rattling quietly northward.

Churchill

Time is necessarily elastic when one takes the train from Winnipeg to Churchill.

It isn't that the distance is especially immense: approximately 625 miles as the crow flies, somewhat longer when following the route of the railway, which veers first to the northwest and then northeast before ultimately turning due north. But the ground, often boggy, is less than ideal terrain; it freezes and melts, stretching and sinking beneath the tracks, which twist and warp, ensuring that for long stretches the train can do little more than crawl.

Accordingly, a journey that by air takes but a couple of hours stretches into approximately two days by rail. Although an official timetable optimistically predicts arrivals at and departures from the various stops en route, it functions primarily as a guide to how far behind schedule the train is falling.

On such a voyage, the key to contentment is a surrender of control, an ability that does not come naturally to all and on this particular occasion seemed especially vexing for one passenger. He studied the Timetable of Questionable Accuracy with the kind of fervor

normally reserved for the Torah, using it — and, perhaps, an unseen
set of actuarial tables — to calculate and loudly pronounce, with each
stop the train made en route, our average speed and likely time of
arrival. Concerned over our glacial progress, he grilled a fellow passenger, a Churchill veteran, for keys to maximizing his time when
eventually we reached our destination. How long would he need to set
aside to be guaranteed a bear sighting? We were scheduled to arrive
on a Saturday afternoon; if he had "done" the polar bears by the end
of Sunday, what would be the best way for him to spend the remaining day before his departure?

Apparently satisfied by the answers, he retreated to his room; his
fellow passenger and I sat in silence, gazing out the windows of the
observation car as the Manitoba wilderness rumbled past, before I began what I hoped was a gentler interrogation.

Doug Ross had until recently been director of the Winnipeg Zoo,
which as we spoke that evening remained home to Debbie, now deceased but then, at age forty-two, the world's oldest known polar bear.
The previous year, Doug had taken early retirement and begun a life
in which, as he put it, he would work only when he needed to, doing
only the things he wanted to. Having spent many years taking care of,
among other charges, a polar bear in captivity, one of the things that
he wanted to do was spend time with polar bears in the wild. So he
had become a driver of a Tundra Buggy, a tourist vehicle that looks
like a bus on monster truck wheels, from which visitors to Churchill
can observe polar bears in relative comfort and safety.

I had spent plenty of time in northern communities, had lived in
a cabin in Alaska, was perfectly accustomed to human communities living cheek by jowl with large wild mammals. But Churchill lies
directly in the path of polar bears that migrate from summer shelter to the coast of Hudson Bay in anticipation of the bay's waters
freezing; it was one thing to have experienced, as I had on several

occasions, a moose lying in the driveway or walking down the street, but the prospect of a large and hungry carnivore lurking around the corner seemed to me an entirely different proposition.

"The thing is," Doug began, "a community like this, we might think it's a big deal to live with polar bears, but for them, it is what it is. I would see kids out in the playgrounds at seven or eight in the evening, and I'd think, 'Wow, is that safe?' But what else are they going to do? I imagine they're taught to be really careful."

We paused as we looked out the window at a brief sign of life, the streets and signposts of an anonymous hamlet past which the train slowly rumbled.

"There were times in the night," he continued, "when I would hear these big crackers going off — you know they fire cracker shells [12-gauge cartridges that explode in the air with a loud bang] from shotguns to scare the bears. Well, first you'd hear all the dogs in town going crazy, and then you'd hear the bangers going off. They tell you all these stories — I don't know if it's just to get us going or what, but they say that, especially if you're out at night, always be sure you know where the nearest car is. Because people leave the doors to their cars and trucks unlocked, just in case."

We headed down to the dining car, and after a light dinner of grilled cheese sandwich and soup, I retired to my accommodation. Officially called a roomette, it could charitably be described as snug. With the bed folded down from the wall, there was enough room for . . . well, the bed. Lifting the bed back up again required stepping out into the hallway; completing the operation revealed a couch of sorts, a padded seat that covered a flush toilet, a fold-down washbasin, and an assortment of cubbyholes for assorted storage. In Manhattan, it probably would rent for about $2,000 a month.

I woke up a few times in the night, as was perhaps to be expected given that I was sleeping in a tiny cabin on a gently rocking train, and

I looked out to see the tall prairie grass and, later, larch and spruce bathed in our transport's reflected glow. The sky was clear, and Orion was instantly recognizable, low on the horizon, as I peeked out my window.

When I slept, I dreamed. Disconcertingly, I dreamed that a polar bear had clambered aboard the train and was at this moment padding in a predatory fashion among the alleyways. The Dream Conductor's assurances that we were all safe did nothing to assuage the fears of Dream Kieran, as he cowered behind his closed and locked door.

The first people to arrive in the environs of what is now Churchill, ancestors of the Inuit who inhabit the region today, did so, as far as we know, approximately three and a half thousand years ago. They were ultimately joined by Chipewyan people from the north and Cree from the south, but a further millennium would elapse before Europeans penetrated the waters of Hudson Bay. The initial wave was English, led by the man after whom the bay is now named, but the first to step ashore and winter by the banks of the Churchill River were Dano-Norwegian.

By the time the *Unicorn* and the *Lamprey*, sailing under the command of Jens Munk, arrived at the mouth of the river in September 1619, their crews were exhausted from a long and trying voyage, punctuated by a battering from severe fall storms. Spying an abundance of small plants—and, in particular, a riot of cranberries—on land, the men anchored their vessels and moved ashore, warming themselves by a fire they maintained day and night, and strengthening from the addition of fresh fruit to their diet.

There was fresh meat, too, to satisfy the men's hunger and improve their health—ptarmigan, caribou, reindeer, and, for reasons the sailors likely neither knew nor particularly dwelled upon, a high concentration of polar bears. But the beneficial bounty didn't last long.

With the onset of winter, the caribou retreated inland, where the snow was deep and soft and where the poorly equipped mariners could not follow. The crew's clothing was inadequate for the cold temperatures and angry winds that tore across the ice of Hudson Bay. By January of 1620, the men began, one by one, to die—not of the cold, but of some unknown disease that made them nauseous, rendered them weak, led to dysentery, and caused them to lose their appetite, further weakening them. By March, the hardened ground and the weakened men combined to make further burial of corpses impossible; the dead were either thrown overboard onto the ice or left on the ground where they had fallen.

In the middle of April, Munk recorded that just five men remained; shortly thereafter, that number had dwindled to three. It would dwindle no further. The return of spring heralded the reemergence of plant life, the juices of which Munk and his companions sucked desperately from their stems. As the ice broke up, the survivors set nets to catch fish; the men's teeth were too loose in their gums for them to chew, so they boiled the fish and drank the broth. When they regained sufficient strength to discharge their ammunition, they did the same with the migrating birds that had returned from the south. Remarkably, on June 16, the three men sailed the *Lamprey* out of Churchill River, into Hudson Bay, and on a successful voyage back to Norway.

For centuries, the prevailing assumption was that the crews' sickness was the result of scurvy; today, another theory holds sway, that the bounty of bear meat that must initially have brought great joy was ultimately the bearer, as it were, of disease and death. Polar bear meat is frequently infected with larvae of the *Trichinella spiralis* roundworm, the cause of trichinosis, a scourge of those unwise enough to eat undercooked game meat or pork. Munk suffered less because he roasted his personal supply more effectively than did his crew, who ate theirs parboiled in vinegar. But partly boiled bear meat is at best an acquired taste; the two men who chose to acquire as little of it as possible were the two who survived to sail the *Lamprey* back to Europe with their captain.

Munk had hoped to make contact with some of the natives he assumed lived in the area, and he rued the fact that his men shot a domesticated dog in the dark of night, thinking it to be a wolf. Had they captured rather than killed it, he reasoned, perhaps they could have tied some trade items around its neck and sent it back home, in the hope that its owners would want to barter food for goods.

In the event, Munk saw not one human denizen during his stay in the area. But it was to do business with those native inhabitants that the Hudson's Bay Company—which in 1670 had been granted a monopoly by England's King Charles II on the trading of furs in the region—established, in 1717, its northernmost outpost, a log fort named after the company's former governor, the first Duke of Marlborough, John Churchill.

In response to growing tensions between England and France over fur trade profits and North American territory, the Hudson's Bay Company in 1732 began construction of a massive stone structure, Fort Prince of Wales, to replace the original log outpost. Like many ambitious building projects before and since, it cost much more and took far longer to finish than initially estimated; not until 1771 was it considered completed. A little over a decade later, three French warships steamed into the harbor and the complement of twenty-two Englishmen manning the fort surrendered without a shot being fired.

Over time, the fur trade declined, and Churchill with it, although a settlement remained. In the late nineteenth century, the members of that settlement turned to whaling, targeting the belugas that populate the bay and profiting from their oil. Between the First and Second World Wars, Churchill bloomed; selected as the site for a new northern harbor, it was connected by rail to Winnipeg to the south and became the portal through which much of Canada's grain was exported. During World War II, the United States Army established a base, Fort Churchill, five miles to the settlement's east; in the years after the war, it functioned as a joint United States–Canadian center for training and experimentation, from which more than 3,500 rockets were launched to study the ionosphere and the aurora borealis that flickered overhead in the night sky.

These were Churchill's glory days, when more than 4,000 people lived at Fort Churchill and in town and the community (relatively speaking) pulsed with energy. But the U.S. Army left the base in 1970, it was used only sporadically in subsequent years, and by 1990 it had been abandoned completely. Today, Churchill's year-round population is approximately 800.

A few buildings still stand as evidence of the base's presence, as well as an airport that is serviced by a runway of a size wholly disproportionate to the population it serves. The port remains active; indeed, Churchill effectively functions as the gateway to Canada's Arctic communities. In fact, although the port is completely ice-free for only three months a year, it handles more than 500,000 tons of grain annually, as well as fuel oil and bulk cargo.

There is talk of revitalizing the community through mining—the town of Thomson to the south has exploded in size and wealth in recent years following development of a nickel mine. There is tourism, too, centered around the town's wildlife: more than 200 species of migratory birds in the spring, belugas in the summer. But it is for six to eight weeks in October and November that Churchill truly comes into its own, when the 800 regulars are joined by approximately 12,000 visitors and seasonal workers, the reason for their arrival evident in the names of the places they eat, sleep, and shop: the Bear Country Inn, the Lazy Bear Café, Great White Bear Tours Gift Shop.

This is bear season, when Churchill, Manitoba, turns the potential, very obvious, negatives of being in the path of the world's largest

carnivore into a lucrative positive, when it revels in its status as the
self-described but undisputed "Polar Bear Capital of the World."

"In Winnipeg, they say you can tell a person from Churchill because
they always look carefully before walking around a corner," chuckled
Tony Bembridge, a man quiet in both demeanor and locution who
for a few months a year interrupts his retirement to help run Hudson
Bay Helicopters for its owner, his son. The company is an important
part of the community. It provides tourist trips of either an hour or
half-hour duration to see polar bears and other wildlife; it assists in
identifying polar bears that are either approaching or have already
entered the town's perimeter; and it provides air transport for
researchers who tag and study Hudson Bay's bears.

One of the company's pilots, twenty-two years old and boasting the
perfect Hollywood pilot name of Jon Talon, sat next to me as Tony
stood up to talk to some prospective customers, fellow refugees from
the Winnipeg train, who smiled and nodded in recognition.

Like anybody who has lived for any time in Churchill, Jon has bear
stories.

"My girlfriend was getting frustrated that after working in town for
seven months she hadn't seen a bear," he began. "'Just be patient,' I
told her. 'Wait till bear season, you'll see one.' Sure enough, we're in
bed one night, when at like four a.m., she wakes up and says, 'What's
that noise? What's what noise?' She throws open the drapes, and there
it is, a bear right outside the window. She starts shouting, 'There's a
bear! There's a bear!' And I start telling her to get dressed because if
that bear comes through that window, we're going out this door."

Bundled up against the buffeting wind that hurled itself across the
bay, Tony and I made our way across the street to Gypsy's Café, a
Churchill institution and gathering place and purveyor of a surprisingly hearty and flavorful French onion soup. We sat at the table nearest the counter, reserved for residents to sit and exchange stories that

in any other locale would seem outlandish and even here sometimes stretch credulity sufficiently to earn the table its own epithet: the Bullshit Table.

Next to us sat Bill Callahan, American by birth but a resident of Churchill for twenty-eight years.

He, too, had a bear story.

Evidently, a community of 800 people is, for Bill, somewhat suffocating, so he lives in a cabin outside of town. It makes for plenty of peace and quiet; but, he says, "I sometimes get some interesting visitors."

One night the previous year, a sow with cubs had pushed through his front door and entered his kitchen while he slept. Placing her paw on the stove in an apparent attempt to reach a loaf of bread that was above it, the sow pressed the button that lit the burner, singed her paw, recoiled, banged into the wall, and crashed out through the now-open doorway, cubs in tow. Having somehow dozed through the breaking down of his door and the presence of three polar bears in his kitchen, Bill was awakened by the sound of the sow thumping into the kitchen wall. Fully naked but half-conscious, he stood in the kitchen doorway, the chaos not yet fully apparent to him, the scene lit only by the glow from the stove, prompting Bill initially to wonder how he could have gone to bed and left the gas flame burning.

Another bear story.

Also from the previous year, this story was told to me in the warmth of a basement room at the south end of town.

As all should, the basement had a glowing fire, a big-screen TV, a wet bar, and a hyperactive dachshund called Monty. Here, too, sat my hosts, Lance and Irene Duncan, proprietors of a bed-and-breakfast in which, Lance advised me sternly after he and Irene had collected me from the railway station, guests are expected to obey two rules: "Please leave a note in our guest book before you leave, and treat our house as your house."

I needed little encouragement to do either, warming myself externally by the fire and internally with the glass of rye I cradled in my hand, as Lance related his close encounter twelve months previously. Both Lance and Irene work full-time jobs in addition to taking care of their guests, Irene at the Seaport Hotel on Kelsey Boulevard (the town's main thoroughfare), Lance, at the time of my visit, at the Churchill Marine Tank Farm by the harbor. It was at the latter, while kicking through the snow in search of a dropped tool at the rail loading area, that Lance rounded a rail car and came face-to-face with a polar bear.

The encounter appeared to startle each equally, the yell that Lance instinctively let out causing the bear to momentarily retreat and buying Lance the time he needed. He leaped into his nearby truck, the door unlocked as always (yes, he said, it is true that Churchill residents leave not just car and truck doors but also house doors open for just such an occasion), and promptly ran the bear out of town, all the way to Cape Merry several miles to the northwest, the site of Jens Munk's enforced overwintering in 1619 and of the fort that the French had captured so effortlessly the following century.

Once he recovered from the initial shock—or, as he jokes now, "once I cleaned out my drawers"—he was, he recalled, mad.

"I was mad at that bear for being there, for giving me a fright, and for costing me part of a morning's work," he explained. "And I was mad at myself for not paying attention and for putting myself in that position."

For Churchill residents, particularly those who, like Lance, grew up in the community, bear awareness is both ingrained and a matter of pride; appropriately safe behaviors are second-nature. The approach is one of neither blustering bravado nor crippling caution; common sense prevails, as does a collective desire to avoid placing human or bear life at unnecessary risk.

"Gone are those days when, if you were a bear that walked into the community, you were a dead bear," mayor Mike Spence told me over coffee at the Seaport Hotel as the wind howled outside. "It was common to shoot twenty-five bears in a bear season."

Such a trigger-happy approach did not translate into greater human safety. In the late sixties, four maulings, one of them fatal, prompted the Manitoba Department of Conservation to post a wildlife officer in town on a permanent basis, but the approach to problem bears remained the same: chase them away if possible; if not possible or if they return, shoot them. The International Fund for Animal Welfare provided funding for some bears to be transported by helicopter to a spot about thirty miles northwest, but beyond that there were simply no other obvious practical alternatives.

That changed with the opening in 1982 of the holding facility, popularly referred to as the Polar Bear Jail. The departure of the military led to a quieter community, which in turn prompted more bears to approach town, thus increasing the likelihood of future encounters. At the same time, a nascent tourist industry was beginning to put greater value on live polar bears, and there were some concerns that the level of shooting might be impacting bear numbers. The holding facility, fashioned from an old hangar at Fort Churchill, provided an alternative that today is an important element of the Polar Bear Alert Program, which seeks to keep bears and humans apart for their mutual benefit.

Signs all over town, on restaurant and shop doors and on lampposts, remind visitors and residents alike of Polar Bear Alert, of the need to exercise caution and awareness, and of the twenty-four-hour "bear line" — 675-BEAR — to report bear sightings.

First line of defense is a series of culvert traps, baited with seal meat, around the perimeter of the community, along the coastline of Hudson Bay and the banks of the Churchill River, the areas along which bears that might pass through town will most commonly traverse. If a bear is caught in a trap, it is taken by truck to the facility, where it will be held for thirty days, drugged, and lifted by Hudson Bay Helicopters to a spot thirty miles or so out of town—or, if it is sufficiently late in the season, onto the sea ice. The facility's purpose is not to punish or deter, but merely to remove an offending bear from the population, to ensure that for the better part of the six-week season it is either in confinement or far enough away from the community that its return is unlikely.

Not all intruders are subjected to a spell in the pokey. Should a bear not be tempted by the culvert traps and manage to make its way into town, said Shaun Bobier, the Manitoba Conservation official in charge of the Polar Bear Alert program at the time of my visit, "we just want to get it off the streets."

To that end, Bobier and his staff employ cracker shells to scare a bear away and then follow it in vehicles to ensure that the interloper leaves town. If, on the other hand, a bear is on the outskirts of the community, or if it returns after being shooed away, the jail beckons.

"We had bears in the back of the Northern Nights Restaurant," said Bobier. "There's huge big willow bushes in the back of there. At some point we might get a bear, he'll come into town, we'll chase it and it goes into those willows, it comes back a couple of hours later that same night. Then we get a couple of reports during the night about the same bear. So what happens is the next morning we'll go to Hudson Bay Helicopters, jump in a helicopter, go to that part of town, dart the bear, and take it to the holding facility."

The polar bears that wander through Churchill are mostly trying to get from A to B, wanting to reach the ice rather than looking for a free lunch. Those that are tempted to tarry by the smell of food (indeed, the majority of all bears that pass through town) tend to be immature bears, subadults that have yet to learn that life is generally easier if Churchill is avoided completely, and for which the pursuit of food, even after the bay freezes and seals are available, remains a sufficiently uncertain endeavor that any supplementary nutrition is eagerly and willingly consumed.

Accordingly, the Polar Bear Patrol was far busier in the days when the town's garbage was piled up and burned in an open-air dump.

"The dump operated up to 2005, and obviously the scents and smells of a garbage dump are going to be a major bear attractant, so what we did, any time there was a buildup of bears, we always had traps at the dump," said Bobier. "We didn't want the bears to become habituated to human food, so we would have to go in and actively remove the bears. Since the dump has closed, we've probably eliminated 60 percent of our bear encounters. Huge, huge drop in the number of bears we've had to handle. And also a huge cost reduction for us, if you think that's an extra fifty, sixty bears we don't have to handle.

"In 2003, the guys handled 173 bears, and at that time we only had 27 holding cells in the facility. But since the dump closed, we've been handling on average sixty or so bears a year. But whether that's a number that's going to remain stable or whether we've just had a couple of nice slow years, I don't know. Next year could be totally different."

In 2007, the year before my visit, 247 calls were made to the bear line, slightly higher than the ten-year average of 221. Of those, Manitoba Conservation staff had to handle (that is, physically capture and/or take to the holding facility) only 49 bears, the fewest in over a decade and barely over half the average of 96. The year with the fewest number of calls prior to my visit was 2005, with just 132 calls — the result, explained Bobier, of ice patterns that year.

"In 2005, the last of the ice came ashore way east in Manitoba and Ontario. So it took the bears a long time to make it up this way, so we had a relatively quiet year," he said.

When we talked in his office in late October, 2008 also was proving to be relatively uneventful, partly because of the ice, although this time for different reasons. The ice had broken up three weeks later than the year before and had come ashore close to Churchill; as a consequence, the bears were fat and contented, in no particular hurry to move anywhere. Plus, it was warm: there was not a hint of snow, and the thermometer had yet to hit freezing.

Conditions would change soon enough, and when they did, said Bobier, so would the number of bears passing through town. Almost half of all calls to the bear line every year are in November, when temperatures start to fall and bears begin to stir in anticipation of being able to set out onto the sea ice; the peak is concentrated over just a few days, when the waters are freezing rapidly and the bears are anxious to begin hunting seals.

"Just prior to freeze-up we have a major spike in activity," Bobier said. "Depending on the season, as long as things stay cold, it'll last four or five days. If we get a sudden temperature rise it may extend it a little. We try not to capture polar bears in those last four or five days; we just try and chase them through."

Since the Polar Bear Alert program began, there has been just one fatality in Churchill, in 1983, and that was a consequence of a perfect storm of unfortunate occurrences. The victim, a resident by the name of Tommy Mutanen, had been rummaging through the freezer of a badly fire-damaged motel and was heading off triumphantly, meat stuffed in his pockets, when he rounded a corner and bumped into a bear. Unable to effect a rapid escape because he walked with crutches, he further exacerbated his plight by attempting to use those crutches in self-defense. Nearby residents heard his screams as he was attacked; despite their best efforts, they were unable to stop the assault and were forced to shoot the bear, by which time the man was dead.

Bobier and colleagues give annual talks on safety to the town's children, lessons that they hope will stay with them through adulthood. Those talks often take place shortly prior to Halloween, when Hudson

Bay Helicopters flies patrols to confirm there are no potential problem bears on the community perimeter, and most available vehicles
are deployed around the edge of town to keep guard while the town's
youngsters (dressed in just about any fashion other than as seals) trick-
or-treat from door to door.

But even the most educated, bear-savvy, cautious populace in the
world can never operate completely free of risk, as long as it shares
space with a predator that is smart, swift, and almost paranormally
quiet. A couple of months before my arrival, Réné Preteau, an em-
ployee at the Northern Studies Center, a research and education complex east of town, was working outside when he turned around and
there, beside him, was a female with two cubs. She raised her paw to
swat at him; he yelled and instinctively struck her with the wrench he
had in his hand.

"I talked to him afterward," said Mike Spence. "He said, 'I thought
I was dead. The thing that saved me was, one of the cubs made a
noise.' She turned, and she went to her cub, and so he ran for the
door, and he said he couldn't remember in the rush of adrenaline
whether to push the door or pull the door. He pulled the door and it
opened, and he said, 'She was so close to me, it hit the bear as I was
pulling it.'"

If some encounters are unavoidable, at least residents know enough
to greatly limit the likelihood of stumbling into those that can be
avoided.

"The community's quite educated on that. They know there's certain areas where you shouldn't be walking," says Spence. "It's generally our visitors who tend to not respect that."

Bobier has been known to refer to tourists as "walking snacks"; the
(at least occasional) validity of the descriptor was highlighted by the
case of the Man with the Walker, clicking forward deliberately, deter-
minedly, and directly for the polar bear that peered at him from behind a rock with ever-increasing intensity.

Fortunately, Bobier arrived before there was bloodshed.

"You do know that bear will kill you?" spluttered the bemused officer.

"If it gets me, it gets me," grumbled the old man.

Another example.

The main town of Churchill is but a handful of blocks north to south, bordered on the west by the river, the railway, and Kelsey Boulevard. The last street before heading out of town is Button Street; it is on this street that Lance and Irene Duncan live, and it was from this street that Lance was about to turn right one evening when he looked to his left and saw three pedestrians heading south, "passing under the last streetlight." Yanking the wheel to the left, he pulled up alongside them and asked what they were doing.

"They said they were heading out of town, away from the streetlights, to get a better view of the northern lights," he recalled. "And I said, well, that's very dangerous, you should stay on the main roads where it is well lit. And at that moment, not forty, fifty feet away, boom, a cracker shell went off. I said, 'Get in the truck,' and they go, 'What's going on?' and I said, 'There's a bear. Just get in the truck.'"

One evening we, too, sought to evade the lights of Churchill — such as they are — to better appreciate the lights in the sky. They were flickering gently as we returned from drinks and dinner at the Seaport Lounge, and Lance steered the truck up the hill and away from Kelsey Boulevard as we sought a better vantage point. We stopped by the cemetery, near the edge of Hudson Bay, the engine running as we peered through the windows that we had cracked open to facilitate our view of the shimmering curtains spreading across the starry night. When we had had our fill and turned to leave, the truck's headlights illuminated a ghostly apparition staring skyward at the cemetery's edge.

We pulled up to the man, standing by his bicycle and drinking in the spectacle, the hood on his overcoat pulled up to protect him from the cold and simultaneously sheltering most of his face from our view.

"Good evening," Lance began. "Are you visiting us?"

Yes, confirmed the man, he was indeed a visitor.

"You know, this is extremely unwise."

The man looked blank, uncertain of the words' meaning.

"This is prime polar bear habitat, and this is peak bear season. We are right by the coast."

The warning did not appear to register.

"You should have seen the lights about a half-hour ago," said the man. "They were incredible."

"That may very well be so," Lance responded, "but it is really incredibly dangerous to stand out here at night. Bears come right by here all the time."

The man turned away wordlessly and slowly walked away, bike at his side, into the darkness.

Before the bears, they came for the birds.

Today, perhaps 500 or so of the 12,000 tourists who descend on Churchill come to gaze at its ornithological attractions in the spring, a number dwarfed by those attracted by the prospect of glimpsing *Ursus maritimus*. But polar bear tourism wasn't even in its infancy, hadn't even been conceived of as a profit-making venture, when some visitors from Texas, in town for the bird watching, asked Len Smith about the polar bears.

Smith, who rented out vehicles to the birders who passed through the community in spring and summer, had often contemplated heading out onto the tundra to watch the bears. But there was a problem: there was no commercially available vehicle that could easily navigate the undulating, treacherous terrain. So Smith decided to build one.

He took the chassis of a Ford F-750 truck, added four big wheels from a crop sprayer, built his own body, and set out. He chuckles now at the memory of some of the things he did in those early days, how on his first couple of trips he didn't even think to take a radio, so when the axle broke, he had no option but to walk back into town to get some help.

At first, he went out there by himself. Then along came a photographer called Dan Guravich, who approached Smith about using his vehicle to make longer trips.

"We would spend two weeks with [the bears] at a time," he recalled. "We would get to know them, and we would name them, and you sure could tell the different personalities."

At first, unused to the intrusion, many bears were uncertain how to react.

"They were skittish, but you took your time," said Smith. "You saw the bear and you let the bear come to you, you didn't go chasing after it. We'd stop there and we'd eat some lunch—once we'd got upwind from them—and slowly, slowly, they would come to you, and as years went on they got to the point where they were comfortable with you. But you got to know them, and it's like with dogs—some dogs you trust, and some you don't. It was the same with bears. But there was never a time when they were aggressive, or when they charged a buggy or anything like that.

"Each bear is different. Sometimes, you might make a little move and the bear will just take off running, and another time you might get right up to it, and it'll be watching you, watching you, but it wouldn't run away."

Then the National Geographic Society came to town, to film a one-hour television documentary about the uneasy relationship between the community's people and its polar bears. The filmmakers included a sequence of Smith and his buggy; once the documentary aired in 1982, the secret was out, and it seemed as if everyone wanted in—first the group from Texas, and then photographers from just about everywhere, so many, in fact, that Smith's initial buggy proved inadequate.

"The first year, I took a group out for five days, and then the next year they doubled it and another guy came in for a week, and from there it just started getting bigger and bigger and I couldn't build buggies fast enough," he remembered.

"Because then once it got out, just about every photographer came up to take pictures that ended up in magazines, then CBC came up to do a one-hour special; it just went crazy. The in-flight magazine for American Airlines, they never even flew into Canada, and I asked, 'What are you guys doing up here?' and they said, 'It's a good story.' Then the Japanese started coming; they wanted to rent buggies for the whole season and bring planeloads of people in and I said, 'No, I can't do that; everyone who started out with me, they get first choice.' Plus, you didn't want to put all your eggs in one basket anyway."

Smith was no longer building buggies around pickup trucks; now he had graduated to larger goods truck frames, to which he added the wider axles of airport fire trucks. Much larger tires not only enabled the vehicles to traverse the terrain but also, at up to five feet high, helped keep curious polar bears at paw's length. The basic design of the body, though tweaked over the years, remains the same: a flat front with a large windshield, buslike windows along the sides and buslike seating inside, and a viewing deck on the back.

The sudden popularity of polar bear tourism threatened to overwhelm the tiny town.

"We had a hard time finding rooms," Smith remembered. "The nice thing about it was the town kind of grew with the business, so as I built more buggies, there'd be another hotel. There were a lot of private homes that would take people in."

On the tundra, meanwhile, the situation was becoming, as Smith puts it, a "zoo." Others followed his lead, and at one point four different companies were making vehicles like Smith's original buggies, all of them crawling over the tundra in search of polar bears. As Smith concedes, they also baited the bears, throwing out blocks of lard as food to entice the fat-loving predators within camera range.

That no longer happens. The operators agreed among themselves to bring the practice to an end, and today feeding a bear, or even tempting one with the aroma of food on the back deck, is strictly forbidden. More than one buggy driver has stated that they either have turned their vehicle around after catching someone violating the rule, or that they unhesitatingly would do so. Vehicles now are licensed, the numbers strictly capped and divided between the two remaining companies. No more than eighteen are allowed to travel — along preexisting trails, carved out in years past by military traffic — in the tundra area close to shore in what is now the Churchill Wildlife Management Area east of town. Twelve of those licenses are assigned to Tundra Buggy, Len Smith's original operation, which also has exclusive access to Cape Churchill, the area in which mature polar bears gather en masse as they await the freezing of the waters.

The other six spots are taken by Great White Bear Tours, whose vehicles, based on airport fire truck frames, are six-wheelers (all except one of the Tundra Buggies have four wheels), larger than those of their rivals, and dubbed Polar Rovers. The company was founded by Don and Marilyn Walkoski in 1994; Don still builds the machines himself with the assistance of a small team of mechanics. He prides himself on the small, civilizing touches that separate his vehicles from the others: the viewing grate in the floor of the back deck; the comfortable coach seats with fold-down lunch trays; and, most frequently commented upon, flush toilets rather than honey buckets.

One day, as evening drew in and the first snow of the season began to fall outside, Donnie — as he is known to friends — looked at one of his machines and showed a visitor the rubber framing around its windshield.

Standing next to the visitor was Donnie's sister, Val, who works for her brother as one of Great White Bear's drivers; on a recent day, she had stopped her vehicle so that her passengers could watch a polar bear that was ambling past, when that bear stood up, dug its claws into the rubber, and tugged.

Donnie, who was in the workshop, heard her call over the radio to another driver as her windshield buckled outward.

"Paul, help me!"

In the event, she did not need any assistance, needed only to turn the key and fire up the engine, the sudden noise and vibration enough to jolt the bear into disengaging. The windshield held, Val continued with the tour, and the excited passengers unanimously agreed it was just about the coolest thing they ever had seen.

Donnie sighed.

"So, now I no longer use rubber to seal my windshields. From now on, I'm going to urethane them in."

He smiled a wry smile.

"That's how it is. You think your day's over and then you're up at three a.m. trying to fix a polar bear–damaged windshield."

Unlike Don Walkoski, Len Smith no longer builds and fixes his own machines — no longer lives in Churchill, in fact, or owns the company he started. He sold Tundra Buggy in 1999 and is now happily retired in Florida; but, he says, "I sure never was sorry about living in the Arctic."

"Welcome to Buggy One!" shouted Robert Buchanan above the growling engine as we lurched uncertainly over the tundra.

There was not, at that point, much to feel welcomed to. Smaller than the Tundra Buggies presently in tourist service, Buggy One had no fixed seating but did boast a pair of bunks nestled in an alcove on the starboard side across from the propane stove. There were a couple of places to balance laptops, but otherwise nothing except a few folding chairs, one of which had an uneasy relationship with my rear and an only occasional interest in the floor, neither of which was aided by the board of plywood that was balanced precariously between the chair and the buggy's side, and which treated every undulation in the terrain as an excuse to crash from one to the other.

"I first started coming up here about twenty years ago," said Buchanan, a white-bearded man whose outsized personality and seemingly boundless energy and enthusiasm belie the fact that he is supposed to be enjoying his retirement and are expressed in a voice that might be described as well projected. "I was invited by a guy named Dan Guravich, who was one of the first photographers of polar bears. He would go out to Cape Churchill in November — early November, back then — and set up camp and photograph polar bears. And it was frontierlike. There was an old school bus for a diner, and it was kind of dragged along by a track vehicle, and you slept in the diner, and you basically had one sled of supplies, a sled shack is what it was.

"It was kind of rough, but we didn't really think anything about it, we just had sleeping bags and so on and so forth. We ate pretty good, we drank pretty good, we had a lot of fun.

"It was nothing for the bears to poke their heads through the window while you were sleeping. And it was nothing to have to push the bears back out of the window and out of your sleeping space. I remember waking up with their breath on my face, and it's quite a rude awakening, and the only thing you can do is to chop them on the nose with your fist as fast as you can and as hard as you can, and they will back off. Of course, if you miss the nose and get them in the mouth, you're in dead trouble and they'll yank you out of that window fast. It doesn't take much. You've got to be dead on target."

In 1999, Buchanan was asked to put together a business plan for Polar Bears Alive, an outfit that had been established by his recently deceased friend Guravich and which, to that point, Buchanan recalled, functioned as "essentially a fraternity" for Churchill regulars. One thorough overhaul later, Polar Bears Alive had become Polar Bears International, and Buchanan was its president. Buchanan describes PBI, which is staffed entirely by volunteers, as a "finance and educational arm for scientists in the field" and a "distributional mechanism" for ensuring that the findings of the scientists PBI supports don't gather dust but are disseminated as widely and swiftly as possible.

It is a task that Buchanan facilitates via a plate-spinning system of mutual benefits, as exemplified by the relationship between PBI and Frontiers North Adventures, which bought Tundra Buggy from Len Smith and operates the company still. Buchanan steers his membership to Tundra Buggy, working with zoos to put together tour groups each bear season; in return, Frontiers North blocks off several of its tour dates for PBI.

"We're doing three trips this year at our lodge in concert with Polar Bears International that we're branding as PBI zoological tours," John Gunter, whose parents, Merv and Lynda, founded Frontiers North, told me one morning in Churchill. "We've crafted an itinerary that we think will well showcase PBI's contributions to conservation awareness in Churchill during polar bear season and make those available to guests through zoological institutions. And in concert with PBI, we host PBI's leadership camp. For the first two weeks in October, we hosted forty students from all over the world who traveled up to Churchill and stayed in the lodge and learned about issues affecting polar bears and the north. The goal of the camp is for students to be able to go back to their communities and be Arctic ambassadors."

And then there was the vehicle in which we presently were lurching across the tundra, which had been built from scratch specifically for PBI's use. It might not have looked like much at that moment, but in a matter of days it would be equipped with video cameras and transmission equipment. These would enable Buchanan and B. J. Kirschhoffer, a young Montanan who was Buggy One's driver and custodian, to post online video of the bears and the scientists who study them, to patch those same researchers through in real time into videoconferences with schools and zoos, and to film interviews with media.

"The thing about these bears," Buchanan resumed, "is that they are incredibly smart and incredibly patient. What we learned over the years was that if they saw an opportunity, they were going to take advantage of that opportunity. Sometimes the storage unit and the old school bus would roll apart, so there'd be a foot gap, and there was a bear once that had watched me for days, and we would have conversations. But he was watching my routine and he knew that every morning I would step across and go into the school bus area for breakfast and then one morning he just reached up and grabbed my boot. Very quickly, in like a quarter of a second, he went ten yards in grabbing my boot, and I was very lucky that I had not tied my laces that day and my foot slipped right out of the boot as he grabbed it. I needless to say wet my pants, and it goes to show that these are really cute and furry animals but they'll kill you in a heartbeat if they get an opportunity, even if they're just playing."

At this point B. J. took pity on the fact that I was being pounded in the back by a piece of plywood from which I was apparently too dim-witted to escape and brought the buggy to a halt while he laid it flat on the floor. As he did so, Robert stood up, walked to one corner, and lifted up the floor to reveal a cage that hung underneath.

"We use that to take photographs of bears' faces, because scientists have found that each bear has a unique whisker pattern, like a fingerprint," he said. "Before we built the cage, we tried to take photographs with a camera that was on the end of a long pole that I'd lower down. And there was one bear, I took a couple of photos and the flash went off in his face, and that bear did not like being flashed. It knew exactly where the heartbeat of that camera was and put one claw right on it. Not anything more than that, he didn't try to hit it or anything, he just put that one claw on the lip of the lens—which, as you know, isn't that wide at all—but he just hooked that claw in there and was not going to let go of that camera. I let it down real fast and that paw just followed right along. I pulled it and yanked it and worked on it for fifteen minutes, and he just concentrated on that camera, and I'd just written off that camera, and then he looked up at me as if to smile and then took the paw off, as if he was saying, 'Just don't flash me.' And I said thank you very much, and that I understood. They're very smart. They're just incredibly smart."

Soon.

It would be time soon.

He had expected it to be sooner, had wondered briefly if the opportunity was in fact just around the corner, once the temperature dropped and the snow blanketed the ground. But experience, even the limited experience of his short life, had taught him that fall's start was often a false one. A few cold days might tease, a layer of snow might prompt excited rolling, but then the winds would change, the air would warm, the snow would disappear, the ground would again be revealed, and the prospect of a return to the ice would seem as far away as ever.

So it had been again this time, the emergence from the summer den coinciding with the temperature's steady fall and night's gradual encroachment, the journey toward the coast hastened by anticipation of an end to the fast. There would be a wait, of course, as there always was, but the crystalline appearance of the water, its cool touch against a tentatively dipped paw, had suggested that perhaps the wait might not be too long. Then the wind changed, delaying the day he yearned for, reminding him of the rumbling protests in his stomach. The wind was slicing and too warm; he had no interest in exposing himself to it. Having found a patch of willows, he shuffled as far into its protective embrace as he was able, the fiercest predator in the world covering his

eyes with his front paws as if not seeing the outside world would ne-
gate its existence, leaving only his well-protected rear end exposed to
the unpleasant elements. With only an occasional, investigative sortie,
he had spent the past few days thus, patiently waiting. It came naturally to him now, patience; it was a most useful trait. It enabled him to
wait for hours on end—when once, as a cub, he had not the disposi-
tion to do so—at a breathing hole, lying perfectly still until the precise
moment when the seal emerged and he could strike. It allowed him to
while away the summer months, even as his body ate away at the layer
of fat he had so ravenously built up on the ice. And it granted him the
ability to wait just a few days more, until the chill returned and the ice
on the bay froze thickly enough for him to return.

Now, the wind had dropped and a chill had returned to the air.
Soon, once more, it would be snowing, and surely more heavily this
time than before. He emerged from the bushes, stretched, sniffed the
air.

That smell again.

He recognized it instantly, as he always did now. Those two-legged
creatures were in the vicinity. There were a great many of them not
far from where he stood, living in an immense herd along the coast.
In times past, he would skirt the edges of their dwellings in search
of food; for his trouble, he had once been chased away with flashing
lights and loud explosions, and as he had grown older he had decided
that discretion and avoidance were wiser options.

Not that the creatures seemed in any way harmful. Some of them
spent time here, too, on the tundra, as if also waiting for the water to
freeze. They traveled on what appeared to be noisy icebergs and uttered strange, hushed, and slightly excited noises—noises that seemed
to be all the more hushed, yet all the more excited, whenever he drew
near.

He could see one of the icebergs up ahead now. He might as well
investigate. It could do no harm.

As he drew closer, though, he realized that this one was not especially interesting. He could smell several of the creatures, but only one was in clear view, staring at him as he walked past. He paused briefly, looked up at it, and continued on his way. There were better things to do. He walked on in hopeful search of a snowdrift, a cool patch in which he could lie and wait for the ice to return.

The gently rising sun cast just enough light and shadow to create confusion. Two nights previously, the first snowfall of the season had blanketed Churchill and environs in a thick, soft white shroud; one night later, fierce winds had blown much of that away, leaving only patches off which the low light ricocheted, confusing the scene. We edged along, the scientists on board joking with each other as they sought to differentiate between sheltering bears and snowy, shadowy background.

"What's that over there? A bear or a rock?"

"Geoff, what do I have to do to get you to spot a bear for me?"

"We're going too slow and too low. Get me an A Star and fly me fifty feet above the deck, and I'll find you some."

"The wind blew away all our nice snow."

"I'd be amazed if there aren't some bears hiding in those willows, maybe covered over with some snow. I wouldn't want to get out and walk there, I know that."

"Is that a bear over there, by the beach? It seems to me every morning there's been a bear wandering along the beach."

"Little Bunny Foo-Foo was hanging out at the lodge earlier. Is that her? No, that's a rock."

"You're getting real good at spotting those rocks."

I gazed absent-mindedly out the window, half-listening to the back-and-forth. A flat-screen monitor in the corner displayed the schoolchildren who had gathered to speak with the assembled scientists, whose banter petered out as they sat abreast on Buggy One looking into a camera, the snowy tundra behind them rolling toward the shores of Hudson Bay.

How far can polar bears swim? How do they hold on to the ice? How long do they live? How big do they get? Are they endangered?

It was one thing to ask those questions of a teacher or even a bear expert visiting the school. It was another to be able to ask them of a gathering of such experts while those experts spoke to them, via the miracle of wireless Internet technology, from amid a gathering of wild polar bears.

"Perhaps we'll see one walk past," one of the researchers had suggested, somewhat optimistically, at the beginning of the teleconference, but although B. J. had driven us away from the camp in search of a suitable backdrop, he had been forced to call off the hunt and park us off the trail in order to set up the camera and establish a link by the appointed time. So there was no polar bear in the shot; but off to the side, among the willows, a movement resolved into a bear ambling, ever so slowly, in our direction.

It was moving as polar bears do: seemingly aimlessly, languorously, its head and neck occasionally swaying loosely and slowly from side to side. I put my camera to my eye, but even at the fullest extent of its telephoto lens, the bear was little more than a small white blob barely large enough to occupy the very center of my picture frame. Whispering to myself, I urged it forward as it padded, inch by agonizing inch, toward us.

But however relaxed a polar bear's stride may be, it can effortlessly eat a up a great deal of distance in a surprisingly short space of time, and by the time I had bundled up and slipped onto the back deck, the bear was no more than thirty, forty yards away, casually looking up at me as it advanced. There was no other buggy in the vicinity, and my companions on Buggy One were inside and occupied; there was only me, my rapid breathing and increased heartbeat, and the approaching bear. Like a ghost's, its approach was silent until suddenly it was so close that its head and then just its snow-dappled muzzle filled my viewfinder. I lowered the camera and looked it in the eye as it looked up at me. Only now, with the bear perhaps three feet astern of me and six or seven feet below, as I braced myself to step away from the edge in case the bear decided to stand up on its rear paws and lean on the deck to take what would have been an uncomfortably close look at me, did I finally hear a noise: almost imperceptible, the softest of crunches of snow underfoot. The bear paused, looked at me, and for a moment we were alone, just the two of us, not another living thing in the world. Then it huffed out a short breath, as if indicating that it had sized me up and decided I was not worthy of any more time. With that, it disengaged our mutual gaze and broke the spell, padding across the tundra and away, never looking back.

To me, mention of the word *lodge*, at least in a wilderness context, evokes images of rustic cabins built from sturdy logs, a roaring fire in the corner and a rug on the floor. At first glance, the Tundra Buggy Lodge—five large modular trailers on huge wheels—does not seek to scale such romantic heights. The lodge's external aesthetic screams functional rather than rustic, and understandably so. As a fixture it is best described as semipermanent: occupying essentially the same location, at a spot on the shore of Hudson Bay dubbed Polar Bear Point, for the bulk of each bear season, it must nonetheless be towed there anew every year and thence, for the final couple of weeks of each season, east to Cape Churchill. Cape Churchill is the *ne plus ultra* of polar bear viewing, indisputably the best place in the world—outside of Nikita Ovsyanikov's cabin on Wrangel Island—to observe mature polar bears assembling in anticipation of being able to set out onto the sea ice. It is accordingly highly restricted in its availability. As runner-up spots go, however, Polar Bear Point is in rarefied air, orchestra seating for an annually unfolding ursine drama.

The lodge comprises two bunkhouses; a cozy lounge car that functions primarily as a kind of anteroom where guests pace and sit in a Pavlovian manner in anticipation of the call for breakfast or dinner; and the dining car where said breakfast and dinner are cooked, served, and eagerly consumed. Between each car, viewing decks afford the opportunity to sneak a cigarette or watch whatever wildlife might be in the vicinity, although during my time at the lodge, the arctic air appeared to discourage most of those on board from tarrying on their way from one car to the next. Most guests, it seemed, preferred to do their bear viewing either during the day on board a Tundra Buggy (two of which — one for each bunk car — are at the dedicated service of the lodge) or from the warmth of the lounge.

It was from the lounge that we watched, absorbed, one evening as two subadult males sparred in the light of the setting sun, throwing snow over each other and wrestling. One of the two appeared to be clearly dominant, forcing the other onto its back where it lay, paws waving in the air, as the first bear pinned it to the ground. As we looked on, a third bear emerged from the orange glow, heading toward the sparring pair as if contemplating whether to turn their tussle into a triple-threat match before apparently changing its mind and veering off in another direction.

The show over, I retired to my comfortable and capacious bunk, wrote in my journal, and fell asleep as the car rocked gently and almost imperceptibly in the fierce winds. In the night, I awoke, looked out the window, and saw a polar bear sheltering from the wind, covered with snow except for its head and neck, which peeked out into the harsh elements. I looked some more, squinted, saw the bear's head moving slightly, until my conscious mind began to assert authority and question how a polar bear could bury itself in a few inches of snow cover and why, even if it could do so, it would leave its head uncovered. With that, the apparition vanished, revealing itself as a wind-blown drift, and I fell asleep again.

When next I woke, northern lights dripped from the heavens, as if a razor blade had sliced open the sky and aurora had spilled out. They shimmered for a while as I watched; I briefly contemplated waking some of my fellow guests, as we had promised each other we would. But although I had seen plenty of aurora displays in my time and knew that this was one of them, I no longer had complete confidence in the way my consciousness was translating the images my eyes were sending to my brain. I did not want to wake a dozen people from their slumber only for them to look at a dark piece of sky where my apparition had been.

By the time I had resolved my inner dialogue, the lights had faded and disappeared, and I closed my eyes once more.

The morning revealed a pattern of bear tracks around and beneath the camp and, a couple of hundred yards away, the same two bears lying where they had wrestled away the evening before, resting now in the fresh layer of snow that had lightly dusted them overnight. When they awoke, they stretched and moved toward each other again. And we watched from a nearby buggy as once more they sparred.

As polar bears almost always are, they were silent, their mood and intent conveyed through body language. They began with a kind of ritualized pre-dance, mouths open but tilted downward, heads held low. Only when each had convinced the other of a lack of aggressive intent did they begin, leaping at each other, grappling on their hind legs, nibbling and gnawing on each other, using their giant paws to push and shove each other, then collapsing to the ground and wrestling, one bear once more on its back, the other nipping at it until they squirmed once more to their feet, rose up on their hind legs, and began again. The bear that appeared to be on the receiving end of most of the exchanges sought on occasion to turn away and even run off a short distance but received by way of a response nothing more sympathetic than a bite on the rear and a resumption of the engagement.

Only when both bears had had enough, the physical exertion

causing them to overheat and seek the cooling comfort of the snow, did the engagement temporarily cease. As it did so, a female and cubs appeared over the horizon, the two offspring glued to their mother's side as she marched in the direction of the lodge and then, catching the scent of the males, paused and stood on her hind legs, smelling the air. Apparently satisfied that the sources of the odor posed at worst a mild threat, she continued onward, albeit on a path that took them a greater distance from the males than their original course would have done. Suddenly, the two cubs, perhaps themselves noticing the juvenile males, bolted in the opposite direction; but the mother did not break her stride, did not pause or in any way react to the cubs' fright, and within seconds they had collected their nerve and resumed their position at her side.

The young males in turn showed only a casual interest in the visitors, raising their heads briefly but lacking either the energy or the inclination to pull their entire bodies away from the cool comfort of the snow. A new arrival, however, produced a greater stir. This was an adult male, its gender and maturity clearly evident in its size and its more rounded physique.* His approach caused more apparent concern in the mother, who now led her cubs away from the lodge, past the lounging adolescents, beneath our buggy, and to a safe distance, her pace fractionally more hurried than before, her gaze cast frequently in the direction of the approaching bull.

---

* Theoretically, there are a number of ways to distinguish males from females, at least when mature. Males, as previously noted, have necks that are wider than their heads, creating a funnel shape from shoulders to face, whereas the heads of females are slightly larger than their necks. Males tend to carry scars about their faces, the results of innumerable bouts of playful sparring and more serious battling for mates. And males are simply heftier. But it isn't always quite so easy to place that theory into practice. Partly because of the amount of fur in the way, even close examination of the genitalia of an anesthetized bear isn't always completely reliable. A 1999 study showed that gender was incorrectly identified in 19 of 139 bears killed by Native hunters in Alaska. But there are times, such as this occasion, when a bear is so obviously massive that it is clearly a mature male.

Now, too, the subadult that had appeared the more enthusiastic to spar with his peer and appeared to have gained the upper paw in most of the exchanges rose to his feet and set off eagerly to intercept the interloper; in a more territorial species, his actions might be interpreted as a challenge or a threat, but in this instance his motivations could only be guessed at. Perhaps it was an invitation to spar, perhaps simply an overdose of teenage hubris. Whatever the case, it became immediately apparent that the move had been a mistake.

The two bears had almost touched noses when the younger one froze on the spot. Whatever the means by which he had done so, the older male had evidently conveyed, in a fraction of a second, that the two animals were not in any way peers and that the subadult had clearly overstepped his boundaries. As if at once recognizing the error of his ways, the youngster began to move slowly away, but in so doing contrived to back himself up against the wheels of the lodge; still keeping his face toward the adult, who had barely moved, he painstakingly maneuvered his rump so that it was free of the obstruction and continued to withdraw. Only then, perhaps because the younger bear was no longer trapped and was free to escape, did the adult move after him, pacing forward, obliging the would-be rival to continue backing up with his head in a low, submissive position, his mouth open. Seemingly satisfied that a sufficient degree of obsequiousness had been displayed, the adult now lay down in the snow and began to groom himself. At this, the juvenile turned and started to jog away; but evidently he had, and not for the first time, misread the situation. The adult leaped to his feet, quickly chased the youngster down, bit him on the rear, and forced him to once more turn and demonstrate his fealty as the older bear lay down to clean himself anew. After another few minutes of submission, the adult was apparently satisfied; when the youngster turned to leave a second time, the adult did not protest, licking his fur quietly as the juvenile returned to his patch in the snow and went back to sleep.

We moved on.

The snow was icy and hardened, blown by the harsh winds into tightly packed mounds of crystal sugar. The sun repeatedly threatened to emerge from behind the clouds, and when, at the midpoint of the afternoon, it carried out that threat, its rays reflected sharply off the packets of silvery snow.

We lurched across the tundra until we came upon two bears dozing in the sun by the left side of the trail. The snow appeared disturbed, suggesting they had been sparring. We drew up to them slowly; they looked at us casually. One hauled itself to its feet and wandered over, sniffed the front tire of the buggy, and began to lick and lightly chew it.* It walked in front of us, licked and chewed on another tire, found a patch of perfectly polar bear–size snow, and lay down in the shade cast by the buggy. In due course, the other bear joined it, each bear now resting its head on its paw, their noses close together and their rumps farther apart, forming a V shape in the shadow.

I slipped outside onto the viewing deck, stood at the edge, my hands in my pockets, and gazed at the bears as they dozed. Periodically, one would open its eyes and look directly at me. I wondered what, if anything, it was thinking. Was it completely indifferent to our presence, so inured to buggy traffic that it barely even paid attention? Was it comfortable but wary? Or was it in fact displaying the predatory patience for which polar bears are famed, lying quietly in anticipation of the moment when one of us would lean too far forward and into striking range?

-----

* Who knows why polar bears do the things they do? I wondered if perhaps the tire had picked up some moisture or some salt. Don Moore of the National Zoo, who had been on Buggy One but was not with us on this particular buggy on this particular day, offered that perhaps the bear was simply exhibiting the species' inherent curiosity and was using all its senses to examine what now stood before it.

The bear closed its eyes again.

In the distance, the waters of the bay rippled slightly. I hunkered down into my coat as an angry wind whipped off the tundra. I pressed myself up against the rear of the buggy to protect myself, and then, all at once, the wind died down, and there was silence.

Melt

Evening on Buggy One.

The descending sun angled the last of its light through the windows as we assembled tables and chairs and arranged paper plates of cheese and crackers. It would have required an especially enthusiastic realtor to describe the furnishings even as functional, but compared to the progressively less clement conditions on the other side of the thin walls, as a rapidly strengthening wind buffeted the buggy's exterior, the setting provided a feeling of protective comfort.

Our visitors arrived, a dozen or so, blowing out their cheeks and rubbing their arms as they escaped the elements and entered our tiny sanctuary. Their journey had hardly been expeditionary in nature — a short walk from the Tundra Buggy Lodge, next to which, after a day of lurching across the tundra in vain search of polar bears, we were now docked — but sufficient in the circumstances for the air outside to chill fingers, bite at cheeks, and make the unremarkable surroundings in which they stood a welcome destination.

To the accompaniment of the nibbling of snacks and sipping of the finest boxed wine northern Manitoba had to offer, Robert Bu-

chanan provided a brief tour—if it could be so described, given that
it required no physical movement on anybody's part other than a slight
turning of the head. Here, below the floor, was the cage—"Except
we don't like to call it a cage; we prefer to call it a platform"—from
which it was possible to photograph polar bears closely and safely.
Over there were computer monitors, camera equipment, and wireless
Internet portals, which among them held the promise of making polar
bear research more versatile and comfortable. No longer did watching
bears require endless days shivering on top of an observation tower at
Cape Churchill; the installation of remote cameras meant that at least
some observations could be conducted more comfortably and just as
effectively anywhere from Buggy One to a base in more salubrious
climes.

"What we're looking to understand from this research," Buchanan
continued, "are three things. What is the census of the bears and how
is it changing? What is the movement of these bears? And we use
satellite tracking, radio collars, and other mechanisms to determine
that. And what are the geographic pockets where these bears are going to survive? Because in all honesty, it is our belief—and I pray
every night that I am wrong about this—that this species will not be
here for future generations. What you are seeing here will not be here
in the next century. We can be sad about that, or we can be motivated
to do something about it."

The audience was captive and sympathetic. Buchanan could address his congregation without fear of contradiction or need for expli-
cation. Those assembled understood his meaning and nodded sagely
and sadly in agreement.

"I want you to do me a favor," he continued. "When you're out
there tomorrow" (and here he nodded over his shoulder and out the
window of the buffeted buggy to the tempestuous tundra outside)
"and you look at a polar bear . . ."

"If we see one!" interjected one of the small crowd good-naturedly;

it had not been a successful couple of days for polar bear viewing, and buggy drivers had been reduced to feigning enthusiasm for sightings of ptarmigans and arctic hares.

Buchanan pivoted to provide sympathetic explanation — "The problem is, when the weather's like this, the bears tuck themselves in the willows and stick their big butts out to protect themselves against the wind" — before returning swiftly to his theme.

"When you're out there tomorrow, do me a favor," he repeated. "Look at a polar bear and close your eyes, just for a second, and imagine if we are the last generation to see a polar bear. We can't allow that to happen. We just can't."

The call from the lodge that dinner was ready arrived on cue, Buchanan's words lingering in the air for a second before they were replaced by the sounds of zippers and Velcro fasteners as the assembly readied itself for the brief return journey. The small crowd made its way, one at a time forcing the door into the resistant wind, which announced its presence with an angry, high-pitched howl.

On one level, warm is relative. Cold is a matter of opinion.

To the visitors who shrank into their thick coats as they hustled out of Buggy One and back into the Tundra Buggy Lodge, there was no question. The fierce wind that roared off Hudson Bay was probing for any kind of entrance, any opening that would enable it to slice into exposed or inadequately insulated flesh. Its aggressive onslaught served only to exacerbate the ambient temperature, which nudged slowly downward with each passing day.

It was cold.

But for the polar bears that lay just out of sight, the wind carried a different message. Wrapped in protective layers of blubber and fur, built for comfort in the coldest of conditions, they huddled in the willows, their large rumps turned protectively outward. Hunger pangs were building, the need to hunt was growing, and the wind, far from chilling them to their ursine bones, served only to remind them with its relative mildness that the sea ice they depend on had yet to arrive.

It was warm.

But there is, of course, a more objective measure, and while weather may change from hour to hour and day to day—the mercury falling then rising, the wind rising then falling—over longer time periods, studies of temperature and climate can reveal a bigger picture. And as Buchanan had alluded to, and his visitors understood all too well, as cold as it may have felt to them at this particular time on this particular day, that bigger picture showed the emergence over the past several decades of an unmistakable trend.

The planet generally, the Arctic particularly, and Hudson Bay specifically were growing warmer.

Earth's climate is not, and never has been, constant. It varies from decade to decade, millennium to millennium, eon to eon.

From about 750 to about 580 million years ago, for example, Earth, it seems, was all but covered with glaciers, a phenomenon referred to by some scientists as Snowball Earth. At the other end of the spectrum, from approximately 250 million years ago to roughly 50 million years ago, the planet was considerably warmer than today; on either side of the extinction event that spelled the demise of the dinosaurs, average temperatures were as much as 9°F above contemporary averages.

At various points in its history, from relatively recently to hundreds of millions of years in the past, Earth's climate has been affected to varying degrees and in varying lengths of time by a suite of factors. Volcanic eruptions, if of sufficient quantity or ferocity, can slightly reduce the amount of solar radiation reaching the planet's surface, bringing about a degree (in the metaphorical, not literal, sense) of cooling for a number of years or even decades. Earth's orbit is slightly elongated

rather than perfectly circular, its axis rotates around an imaginary cen-
terline, and its tilt goes up and down, and all three shift on regular cycles; an intersection of two of these — its tilt and its axis — combined
to flood the Northern Hemisphere with solar energy around 15,000
years ago and bring an end to the last Ice Age.

But the most consistent factor in Earth's climate warming or cooling is the composition of its atmospheric gases, a notion that was first
mooted early in the nineteenth century by French scientist Joseph
Fourier. He wondered why, given that solar radiation was constantly
hitting the surface of Earth, the planet didn't keep heating up and ultimately become as hot as the star it orbited. The obvious answer, he
concluded, was that energy was being radiated back out to space; but
when he worked out the arithmetic, the answer to his equations was
an Earth that was below freezing. The explanation, he surmised, was
that the atmosphere was trapping some of that heat — as if, he said,
the planet were a box covered by a pane of glass, through which, as in
a greenhouse, sunlight could enter but heat did not escape.

In 1859, British scientist John Tyndall sought to identify which of
the gases in the atmosphere would be most likely to perform such
a feat. Through a series of tests in his laboratory, he determined
that the atmosphere's primary constituents, oxygen and nitrogen,
were transparent to the sun's infrared radiation and thus not a factor.
Methane, however, was, as Spencer Weart describes it in *The Discovery
of Global Warming*, "as opaque as a plank of wood"; so, too, was a gas
of seemingly little consequence in the atmosphere, carbon dioxide. It
seemed unlikely that carbon dioxide on its own could have much impact on temperatures, reasoned Tyndall, because it constituted such a
small percentage of atmospheric gases. Of likely greater import, he
proposed, was water vapor, which is a more voluminous greenhouse
gas and which, Tyndall poetically declared, "is a blanket more necessary to the vegetable life of England than clothing is to man."

Thirty-seven years later, Swedish scientist Svante Arrhenius realized that a small increase in carbon dioxide might warm up the atmosphere sufficiently to allow it to hold more water vapor—which would in turn lead to further warming. Arrhenius also recognized that such increases in carbon dioxide levels could result from industrial processes, specifically the burning of fossil fuels such as coal and oil. Because those fuels contain carbon, their combustion returns carbon to the atmosphere, where it combines with oxygen to create CO_2. At the time, the amount of carbon that had been released was, he calculated, insufficient to make much difference; but, he reckoned, were the amount of CO_2 in the atmosphere to double, global temperatures could increase by as much as 8°F to 9°F. Based on the emission rate at the time, he estimated that such an eventuality would not unfold for at least two thousand years. A little more than a century later, his calculations are being put to the test far more rapidly than he imagined.

Scientists employ a number of devices to measure past climate and atmospheric conditions. Because, absent the invention of a time machine, these conditions must be inferred, the evidence on which they rely is called proxy data. Those can include anything from historical records—such as diaries and ships' logs—to the distribution of fossil corals and the rings in trees. Of particular value, however, in determining levels of atmospheric carbon dioxide up to 800,000 or so years into the past are ice cores, cylindrical samples gathered by drilling deep into the Greenland and Antarctic ice sheets, where thousands upon thousands of tiny air bubbles have collected over the eons, perfect snapshots of past atmospheric composition.

By analyzing such ice cores, researchers have been able to determine that during the last Ice Age, levels of CO_2 in the atmosphere were approximately 200 parts per million (ppm). In preindustrial times, those levels were closer to 285 ppm. Since the middle of the

eighteenth century, however, and the dawn of the Industrial Revolution, those levels have grown, steadily at first and then with increasing rapidity. In 1958, the first year of a continuous measurement of
atmospheric CO_2 levels from Mauna Loa in Hawaii, that figure had
climbed above 310 ppm. It has increased since, and at the time of this
writing, in late 2009, it is close to 390 ppm. The difference between
now and the period before the Industrial Revolution, in other words,
is greater than the difference between immediately preindustrial times
and the Ice Age.

As a consequence, according to the Intergovernmental Panel on
Climate Change (IPCC), global temperature has increased by an average of 1.3°F over the past century, and average Northern Hemisphere temperatures during the second half of the twentieth century
were very likely greater than in any 50-year period over the past 500,
and quite possibly the last 1,300, years. And global temperatures are
continuing to climb, ever more steeply: the 1980s were the hottest decade on record, until the 1990s, which were, on average .25°F warmer.
The first decade of the twenty-first century was warmer still—.36°F
warmer than the nineties.

The warming that has taken place, and that which is predicted
to, has not been and will not be uniform. Some areas—parts of the
southeastern United States, for example—have, if anything, experienced a slight cooling. It is possible that feedback mechanisms result-
ing from, for example, huge influxes of fresh water into the North
Atlantic as a consequence of melting of the Greenland ice sheet could
disrupt warm water currents and cause western Europe to experience
either cooling or at least reduced warming. Equatorial regions are, in
general, so far showing relatively little temperature increase. But the
regions of the world where warming is already most measurable and
extreme, where impacts are already noticeable and are predicted to
advance most rapidly, are at the ends of the Earth, in the Antarctic
and, especially, the Arctic.

Although here, too, there has been regional variation*—Alaska and Siberia, for example, appear to be warming more rapidly than Greenland—average annual temperatures in the Arctic have increased by approximately 0.75°F since the mid-1960s—more than four times the rate for the globe as a whole and enough for a recent study to conclude that the Arctic is warmer now than it has been for at least two thousand years.

There are several reasons why the Arctic should be at the forefront of global warming. The "weather layer" of the atmosphere known as the troposphere is thinner above the Arctic than over the equator; as a consequence, it takes less energy to create a given amount of warming. In tropical regions, where the air is already warmer and thus more humid, a greater proportion of the sun's energy is expended on evaporation; the drier air of the Arctic offers no such obstacles and so that energy leads directly to heating.

And once the heating begins, it feeds on itself, more heating causing more heating until eventually the cycle is in danger of becoming unstoppable. To understand why, once the Arctic begins warming, it is increasingly and ultimately irresistibly vulnerable to further warming, we must look to the very thing that helps define it, that is both a consequence and facilitator of its normal frigid state.

The white shroud that characterizes polar regions descended upon them because their lower temperatures enabled it to. And then it

---

* In the Antarctic, as well. Although the West Antarctic Ice Sheet has shown signs of warming and melting—particularly the Antarctic Peninsula, where the Larsen A, Larsen B, and Wilkins ice shelves have essentially disintegrated—parts of the larger East Antarctic Ice Sheet have been cooling. Part of the reason for this seems to be the existence of the "hole" in the stratospheric ozone layer above Antarctica, which causes an intensification in cooling wind patterns. Ironically, as the ozone hole repairs itself following international regulations to eliminate the chemicals that created it, that effect will be eliminated and Antarctica is likely to warm more rapidly—the solution to one environmental problem contributing to another.

helped perpetuate those temperatures because ice and snow reflect sunlight back into space, a phenomenon known as albedo. But as the ice begins to melt, the darker land and ocean that emerge from beneath absorb, instead of reflect, the sun's energy. (Whereas ice *reflects* as much as 90 percent of the solar radiation that strikes it, for example, the surface of the ocean *absorbs* a similar percentage.) In the Antarctic, the greatest contributor by far to the albedo effect is the giant ice cap that blankets the continent year-round. At the other end of the globe, surpassing even the Greenland ice sheet, that honor belongs to the sea ice of the Arctic Ocean.*

For comparison, we revisit the Antarctic. Unlike its northern cousin, sea ice in the Antarctic is a purely seasonal affair. Extensive in winter but unrestrained by landmasses, in summer it breaks up and drifts away. The Antarctic winter sea ice extent has not shown any appreciable change — if anything, it has shown a slight, if statistically insignificant, increase; in summer, it melts almost in its entirety as it has always done.

But in the northern realms, sea ice is constrained by the boundaries of the landmasses that encircle the Arctic Ocean. With the exception of the Fram Strait east of Greenland and a smaller artery, the Nares Strait, between Greenland and Ellesmere Island, there is no avenue through which sea ice can escape. As a result, while much Arctic sea ice melts each spring and summer, ice floes in the higher latitudes can persist from year to year, traveling around the Arctic Ocean basin, grinding into and on top of each other, becoming thicker and more resistant to melt. Which makes sea ice in the Arctic a much more interesting barometer than that of the Antarctic; under no scenarios as

---

* There is another nightmare scenario. Locked in ice crystals in Arctic soils and ocean sediments are billions of tons of methane; should warming advance sufficiently to melt those crystals and release that methane — which, as John Tyndall discovered, is itself a highly potent greenhouse gas — the consequences would far outstrip any feedback effects from diminished albedo.

presently imagined would sea ice fail to form in the boreal winter any more than in the austral one. But it stands to reason that, as the Arctic warms, thick ice would become thinner ice as it melts, thinner ice would melt before it has a chance to thicken, the extent and volume of Arctic sea ice would diminish, and the pace of decline would accelerate as temperatures increase and larger areas of heat-absorbing ocean promote further warming.

Which is, it seems, exactly what is happening.

In 1978, NASA launched the Scanning Multichannel Microwave Radiometer (SMMR), which, because it uses microwaves, is able to image the ice cap through clouds and in darkness. Ever since, scientists have been able to map Arctic Ocean sea ice as it contracts and expands, painting a constantly changing picture of its movements and plotting a long-term portrait of its trends. Combined with shipping records and ice charts, those satellite records show unequivocally that Arctic sea ice extent has been declining since at least the 1950s and that, much like the increase in global temperatures that has driven it, that decline has increased sharply in the past decade.

Each September, researchers at the National Snow and Ice Data Center (NSIDC) in Boulder, Colorado, watch the daily images from the SMMR, and two other satellites that have been launched in subsequent years, with particular intensity. Over the course of the summer, the sea ice in the Arctic gradually decreases, until, by roughly the midpoint of the ninth month, the onset of fall arrests its demise. For a while, it may alternate, one day growing a little in extent, the next day shrinking again, responding to variations in the weather as summer fights to keep its last tenuous grip on the Northern Hemisphere. Finally, after the sea ice cover expands clearly, even if slightly, for several consecutive days, NSIDC releases information on the lowest extent it had reached to that point, which it dubs the sea ice minimum.

From the time NSIDC began to monitor SMMR imagery, the overall picture was of a decline in sea ice extent. But even as it trended downward, it did so arrestingly: for the first twelve years, even a poor summer ice year was generally followed by a recovery to average, or near-average, conditions over the winter. That all changed in 2002.

That year, the summer sea ice minimum fell to a record low 2.3 million square miles, more than 400,000 square miles below the 1979–2000 average and almost 66,000 square miles less than the previous record low. The following year, the rebound, such as it was, was a mere 7,700 square miles above the pre-2002 record. In 2005, sea ice extent fell to another low, 2.15 million square miles.

Then, in 2007, a high-pressure system hovered over the central Arctic Ocean for much of the year, opening the sea ice to the sun's rays, which beamed uninterrupted through clear blue skies. Those same atmospheric conditions brought warm winds into the Arctic, prompting further melt and pushing ice away from the Siberian coast. The combination of sunlight and warm winds was too much for an ice cap that was thinner than usual after recovering poorly the previous winter from its September minimum.

The result was a massive drop in sea ice cover, well below even the 2005 level, to 1.65 million square miles, shattering the previous record by 23 percent, some 39 percent below the 1979–2000 average and, reckoned NSIDC, half of what it had been in the 1950s.

The following year was little better, even without the same perfect storm of conditions that had prevailed in 2007. The minimum extent, 1.8 million square miles, was lower than in every year except the previous one, prompting NSIDC's Julienne Stroeve to observe, "I find it incredible that we came so close to beating the 2007 record — without the especially warm and clear conditions we saw last summer. I hate to think what 2008 might have looked like if weather patterns had set up in a more extreme way."

There was a recovery of sorts in 2009, largely because wind patterns spread ice around instead of compacting it as had been the case two years previously. But even so, sea ice extent was the third lowest on record, and of particular concern was the fact that not only was sea ice, on average, far lower in area than it had been only a few years previously, but it was also thinner and younger.

In 1987, 57 percent of the ice pack was at least five years old, and a quarter of that ice was at least nine years old. By 2007, only 7 percent was at least five years old, and virtually none was nine years old or more. In 2009, only 19 percent of the ice cover was over two years old, the least in the satellite record. That same year, researchers from NASA and the University of Washington published satellite data showing that ice thickness had declined by 2.2 feet between 2002 and 2008; ten years earlier, another study had shown that sea ice thickness had already decreased by about 4 feet between the 1950s and the 1990s.

Even as the ice thinned and melted, currents continued to flush some of it south through the Fram Strait, and that ice's removal, combined with the melting of the ice that remained and the diminishing albedo caused by that melting, threatened to push the Arctic ice pack past a tipping point, after which thinning ice would lead to further thinning ice, melting to more melting, until eventually there would be little or no summer sea ice remaining in the Arctic at all.

At the same time NSIDC announced the 2009 sea ice minimum, the *Arctic Sunrise* was grinding its way through the ice edge of the Fram Strait, carrying a team of scientists investigating sea ice thickness and melt rate. In the ship's mess, the research team's chief scientist, Peter Wadhams of the University of Cambridge, briefed the crew in stark, almost apocalyptic, terms.

"We are entering a new epoch of sea ice melt in the Arctic Ocean due to climate change," he said. "In five years' time most of the sea ice could be gone in summer with just an 'Alamo of ice' remaining north of Canada's Ellesmere Island. In twenty years' time, that will also be gone, leaving the Arctic Ocean completely ice-free in summer. In the last few years, there has been an unprecedented retreat of the sea ice in the Arctic Ocean during summer months, but this starts during the winter. So there's a decline in the rate of growth of sea ice during the winter, an increase in the rate of melt in the summer, and now the thickness of Arctic sea ice has diminished much more rapidly than it had in previous decades. At some point, the ice will not grow enough in winter to match summer melting, and the summer ice will disappear, all in one go."

As sea ice disappears, so, too, do the algae that it contains and that support the very foundations of the Arctic marine ecosystem. Because ice floes extrude salt as they form, they are fresher than the ocean that surrounds them; studies in the Beaufort Sea have suggested that, as floes have melted at a greater rate and extent, they have created a layer of relatively fresh surface water fully 100 feet deep, one-third deeper than was the case twenty years ago. As a consequence, some of the larger algae species that provide fuel for the biological engine that drives the Arctic marine ecosystem have diminished, replaced by smaller species more commonly associated with freshwater environments.

Changes in the composition of algae beneath the ice threaten the productivity of the Arctic marine system, as does the disappearance of the ice itself. The ice edge zone, the most biologically productive area in the northern polar ocean, is especially productive close to shore, above the shallow waters atop continental shelves. For walruses, these areas are a veritable smorgasbord, the ice floes on which they rest providing the perfect platform from which to dive to the bottom, where they graze on clams and other shellfish that they suck from the sea floor. As the sea ice retreats from the shore, the ice edge becomes located above deeper waters, forcing walruses to travel progressively longer distances, and expend progressively more energy, to reach their food source — until eventually, if Wadhams's bleak assessment is correct (and there are many others who propose a more conservative timeline, but who nonetheless agree with the notion of partially or wholly ice-free Arctic Ocean summers sometime around mid- or late-century), they will not even have any ice floes from which to launch their assault on benthic bivalves or on which to roam the frigid waters of the north.

Not only that, but in a warming Arctic the benthic buffet may be relatively barren.

"In the Bering Sea, for example, the spring phytoplankton bloom accounts for much of the primary production and makes for a very, very rich ecosystem," says Brendan Kelly of the University of Alaska. "It used to be the case that a lot of the phytoplankton would essentially fall out of the water column to the benthos. And part of the reason that was happening was that bloom happened at a time when the water was cold enough that the zooplankton community couldn't mature fast enough to eat all that phytoplankton before it sank. Hence it was delivered to the bottom, which is why you have this rich benthic community. Now what's happening is that it's becoming warm enough that the zooplankton community can in fact mature and consume most of that production. So now you're shifting the system away from the benthos. So bearded seals are in trouble, walruses are in trouble. But fin whales and humpback whales, the pelagic-feeding consumers, they're going to do great. What we're likely to see is a huge shift in which species are favored and which are disadvantaged."

That assessment was shared by the authors of a 2008 study in the journal *Ecological Applications*; seasonally migrant Arctic species such as fin, minke, humpback, and killer whales, they noted, are likely to find increased opportunities in a warming Arctic, not only because of a shift to a more pelagic ecosystem but also because dwindling sea ice cover would grant them easier access to that ecosystem. But at the other end of the spectrum, they continued, are four "ice-obligate" species that depend on sea ice as a platform for hunting, breeding, and resting, and for which future prospects are dim indeed. They listed the walrus as one of those species; bearded and ringed seals were two of the other three. For ringed seals in particular, the adaptation to ice is profound, an essential element of their very existence. They alone have evolved sharp claws for digging holes through floes, have turned what is to most others an obstacle to be hurdled into a niche to be exploited. Every aspect of their lives is tied to sea ice. They need a stable platform on which to haul out and rear their young in spring and ample snow cover over that platform in order to build lairs in which to give birth to those young. Should the ice break up, or the snow melt, earlier in the year, newborn pups could drift away from their mothers or be left exposed on the surface of the ice, at the mercy of the ringed seal's chief predator.

Such a development would not, however, be to the benefit of the predator, which has less to gain from eating ringed seals when they are vulnerable, newly born, and small, no matter how easy the pickings, than from hunting them when they are vulnerable, newly weaned, and fat. It is to take advantage of the abundance of ringed seal pups in their latter state that the predator's young are born at a particular time of year; in the harsh Arctic environment, a shift of just a few weeks, or a decrease in the number of pups that even make it to the fat, weaned stage, could prove disastrous. For in the same way that ringed seals have evolved specifically to take advantage of the conditions provided by ice floes, so this one predator has evolved specifically to take advantage of ringed seals. That predator, the fourth member of the afore-mentioned "ice-obligate" club, is, of course, the polar bear.

In 1988, when Ian Stirling wrote his definitive monograph on polar bears, he included, naturally enough, a chapter on conservation concerns. The chapter covered the rise and fall of commercial hunting and the development of the polar bear agreement. It considered the possible impact of toxic pollutants such as mercury and PCBs. And it dwelled briefly on issues of habitat modification and disturbance from such activities as seismic testing and oil drilling. As Stirling explained then, and as subsequent studies have served to confirm, the impact of all these on polar bears is uncertain.

Heavy metals such as mercury, and artificial compounds known collectively as persistent organic pollutants (POPs), are carried by winds from the industrialized world into the relatively pristine Arctic realm and deposited by precipitation over the ocean, where they enter the marine food web. Their concentrations become larger as they travel up the food chain, from planktivorous fish to carnivorous fish, to seals and thence to the apex of the Arctic marine ecosystem, the polar bear.

Polar bears in some parts of the Arctic have been found to contain mercury in levels greater than those shown to cause severe neurological ailments and death in humans, but whether such high levels have similar effects in the bears remains unclear. There is some evidence to suggest that polar bears are somehow able to metabolize mercury in their bodies into a relatively harmless form, and thus spare themselves from its worst impacts; contamination by POPs, however, may be another matter. One study has shown a strong correlation in polar bears between high levels of contaminants known as polychlorinated biphenyls (PCBs) and low levels of antibodies, and other studies have suggested the same contaminants may cause reproductive and hormonal abnormalities — such as reduced fertility in females and smaller testes in males.

As the authors of one study laconically observed, however, there remains "a substantial knowledge gap."

Such a gap exists, also, when considering to what extent polar bears are affected by nearby human activities; the particular concern is for those bears that cannot simply move away from any noise or disturbance — specifically denning mothers and cubs.

Although numerous studies of bears in dens have suggested that all but the very closest and loudest of noise sources are of little concern to them, that they show signs of anxiety only on relatively rare occasions and feel impelled to abandon their dens only in situations of extremis, the fact does remain that their immobility creates a degree of vulnerability. There is also consideration that the audio frequencies of industrial activity could interfere with what, to human ears, are unheard vocalizations between mother and cubs, a subject of ongoing research.

But polar bears are also smart and adaptable animals, and there is much circumstantial evidence that, once they have ascertained that an activity is not obviously harmful to them, they tend to simply avoid, ignore, or even exploit it.

That said, to dismiss the potential negative impacts of oil development and industrial activity in polar bear territory would be cavalier, and researchers do not do so. Even if direct impacts of a particular activity may appear to be limited, there are possible secondary considerations: the fact that development often begets development; the prospect that polar bears that become accustomed to human activity may be more likely to have encounters with the humans concerned, almost certainly leading to an unhappy ending for one or both; the potential, in the case of oil exploration and extraction, for a blowout or even a relatively small spill that could contaminate bears directly or via oiling of ice and water and the seals thereon and therein.

Stirling discussed all these issues in his 1988 book, but he did not address what now seems a glaring absence, what is in the early years of the twenty-first century the most widely discussed environmental consequence of drilling for oil and burning fossil fuels: its ultimate impact on the global climate and thus, most important in this particular context, on the Arctic sea ice and the species that depend on that sea ice for their survival. That he did not do so is not surprising. The mechanisms by which greenhouse gas emissions might warm the planet had been long established, of course; observations from

Mauna Loa had been monitoring an increase in atmospheric carbon dioxide for thirty years; the scientists on the leading edge of the issue, epitomized by but by no means limited to NASA's James Hansen, had begun sounding the alarm, their concerns amplified by a few prescient writers. But climate models were in their relative infancy and, compared to those that were to come, were simplistic representations of toy planets; Hansen himself predicted that the clarity of anthropogenic global warming would only truly emerge from the background statistical noise by the turn of the century. And some writers were less prescient than others: in 1987, for example, a young observer of matters environmental named Kieran Mulvaney devoted approximately ten pages of a fifteen-page book chapter to tropical deforestation and only a couple of paragraphs to the emerging issue of climate change, the possible consequences of which he confessed sounded a little "like science-fiction."

But in the years immediately following his book's publication, Stirling noticed something puzzling. The bears he studied year after year in western Hudson Bay, near Churchill, were declining—not in numbers, necessarily, but in size. Their famously rotund profiles were becoming comparatively skinny. Stirling and graduate student Andrew Derocher, who himself has gone on to become one of the leading lights of what remains a fairly small band of polar bear field biologists, looked at the data, crunched them, reexamined them, looked at them from a different angle, and crunched them again; but however they approached the data, the conclusions were the same. It was a linear, one-way trend, which initially prompted the two researchers to discount climate as a possible factor because, they reasoned, something with such annual variability could not account for such a steady progression. They considered other possible factors—pollution, hunting, other forms of human interaction—before returning to the explanation they had earlier set aside.

The result was the 1993 publication in the journal *Arctic* of a pa-

per entitled "Possible impacts of climate warming on polar bears."
The paper did not attempt to make a definitive link between the
decreased size of bears in Hudson Bay and any warming in the climate, but Stirling and Derocher did note that, during 1992 when the
summer was in fact anomalously *colder* than the mean average—likely
because of the effects of the eruption of Mount Pinatubo in the Phil-
ippines—sea ice cover on Hudson Bay lasted an extra three to four
weeks, and the bears not only took advantage of every extra day, they
accordingly returned to land for their summer fast fatter than they
had been for some years. This suggested, they wrote, "that polar bears
remain on the sea ice hunting seals for as long as possible," and that
therefore, conversely, diminished time on the ice as a result of the ice
breaking up earlier in summer or forming later in fall could have an
opposite, deleterious effect. Starting with a formulation by a different
researcher that an increase of just under 1°F in average temperature
in the Hudson Bay region (significantly less than models predicted for
northern Canada) might be sufficient to cause the ice to break up one
week earlier in western Hudson Bay and two weeks earlier in the bay's
eastern region, Stirling and Derocher calculated a dietary intake for
an average female polar bear, as well as the amount of fat burned during the summer fasting period. They concluded that, if ice broke up
one week earlier and refroze one week later, a female bear in Hudson Bay would return to the ice on average about 48.5 pounds lighter.
Were that to be the case, they wrote, "fewer adult female polar bears
would be able to store enough body fat to produce and successfully
wean cubs. Eventually, cub production would not balance mortality
and the population would decline. Furthermore, it is likely that females that were successful at raising cubs to the fall would be unable
to nurse them through the ice-free period because of being unable to
store adequate fat reserves."

The paper's summary was groundbreaking in its conclusion:

If climatic warming occurs, the first impacts on polar bears will be felt at the southern limits of their distribution, such as in James and Hudson bays, where the whole population is already forced to fast for approximately four months when the sea ice melts during the summer . . . Early signs of impact will include declining body condition, lowered reproductive rates, reduced survival of cubs, and an increase in polar bear–human interactions . . . In the High Arctic, a decrease in ice cover may stimulate an initial increase in biological productivity. Eventually however, it is likely that seal populations will decline wherever the quality and availability of breeding habitat are reduced . . . Should the Arctic Ocean become seasonally ice free for a long enough period, it is likely polar bears would become extirpated from at least the southern part of their range.

Six years later, Stirling, writing now with Nick Lunn and John Iacozza, published another paper in *Arctic*. The tone this time was stronger. The additional years of observation had provided the researchers with the confidence to move beyond the purely speculative and theoretical.

"From 1981 through 1998, the condition of adult male and female polar bears has declined significantly in western Hudson Bay, as have natality* and the proportion of yearling cubs caught during the open water period that were independent at the time of capture," they wrote. Over this same period, they continued, the breakup of the sea ice on western Hudson Bay had been occurring earlier; each year, the earlier the ice broke up, the poorer the condition of the bears as they came to shore. They concluded: "The trend toward earlier breakup was also correlated with rising spring air temperatures over the study area from 1950 to 1990 . . . The ultimate factor responsible for the earlier breakup in western Hudson Bay appears to be a long-term warming trend in April–June atmospheric temperatures."

---

* Essentially, the number of live births per capita in the population.

By now, climate change had moved to the front burner scientifically and politically; the signing of the United Nations Framework Convention on Climate Change in 1992 and the addition of the Kyoto Protocol in 1997, which established mandatory limits for greenhouse gas emissions, marked the first tentative forays into international cooperation to address global warming. The formation in 1988 of the Intergovernmental Panel on Climate Change (IPCC) created an international scientific exchange and repository and kick-started a series of approximately five-yearly assessment reports, each updating the one before and each providing the most comprehensive consensus summary of climate-related observations and predictions available at the time.

Regionally, a scientific symposium in Reykjavik, Iceland, in November 2004 saw the presentation of the results of the Arctic Climate Impacts Assessment (ACIA), which were made available, publicly and gratis, in a 140-page synthesis report entitled *Impacts of a Warming Arctic*. The report acknowledged that "Earth's climate is changing," that "these climate changes are being experienced particularly intensely in the Arctic," and that, among the impacts of such changes, "reduction in sea ice is very likely to have devastating consequences for polar bears, ice-dependent seals, and local people for whom these animals are a primary food source." In a press release issued at the conclusion of its eleventh meeting, in Copenhagen, Denmark, in 1993, the IUCN Polar Bear Specialist Group referenced publicly for the first time concern "about the possible detrimental effects of climate warming on polar bears." By the time of its thirteenth meeting, in Nuuk, Greenland, in 2001, the group referred to "ecological change in the Arctic as a result of climate change and pollution" as posing one of the "greatest future challenges to conservation of polar bears."

Three years later, Andy Derocher, Ian Stirling, and Nick Lunn published a synthesis of all the available information on likely consequences for polar bears of a changing Arctic, both observational and conjectural.

They affirmed that changes in sea ice extent and the timing of both freeze-up and breakup pose potentially the most significant and fundamental threat to the existence of polar bears throughout their range by reducing the availability and abundance of seals — although, they pointed out, in the short term there is the prospect that climatic warming might improve bear and seal habitats in some areas if thick multiyear ice were replaced by annual ice with more leads. As sea ice thins, they continued, it would become not only more fractured but also, as a consequence, more likely to be at the mercy of ocean currents, ice islands twisting and turning and circling around Arctic waters. And while polar bears routinely wander vast distances across shifting pack ice, ice that moves farther and more rapidly would transform the slowly rotating gyres on which they routinely travel into more swiftly and unpredictably swirling eddies. This would force the bears to work harder against the ice's direction in order to return — as, for all the distance they traverse, they so often do — to the point from which they came. And they'd be required on occasion to navigate ice that is thin and ill suited to supporting the weight of a large carnivore, forcing them at times even into the water, obliging them to swim for longer than they are accustomed. All of this would combine to sap them of energy in an environment in which energy is always at a premium and in which, with their favored prey likely less numerous and less accessible, it would be even more so.

Not all impacts are created equal, and for pregnant females the consequences of a combination of diminishing, more rapidly moving ice and polar bears' inherent fidelity to certain sites could be particularly severe. Wrote the three researchers: "As the distance increases between the southern edge of the pack ice, where some polar bear populations spend the summer, and coastal areas where pregnant females den, it will become progressively more difficult for them to reach their

presently preferred locations." A previous study, they noted, had suggested that "by the 2050s, the mean minimum extent of the sea ice
in the polar basin would be about 600 km [370 miles] from the north
coast of Alaska or western Siberia and 100 or so km [roughly 60
miles] north of Svalbard. Two of the three largest known polar bear
denning areas are on Wrangel Island and the Svalbard Archipelago.
It seems likely that if this prediction is correct, pregnant females will
likely not be able to reach either of these areas." The prognosis was
no less bleak for those polar bears that choose to den on sea ice, as
do a slight majority of the female bears in the Beaufort Sea region.
The anticipated extra fractures in the ice, and the likelihood of floes
and ice islands drifting farther and faster, raise the prospect that the
distance from core hunting grounds that the dens travel while mother
and cubs are inside, already regularly in the order of hundreds of
miles, will in the future prove far too great for a newly emerged, un-
dernourished female and two tiny youngsters to navigate.

In the end, they wrote, while all bear species are intelligent and
adaptable, polar bears no less than any others, the rapid rate of
change in the Arctic and polar bears' highly specialized nature lead
to one inescapable conclusion: were sea ice to disappear completely
as had been predicted by some, "it is unlikely that polar bears will survive as a species."

There were, and are, dissenters, critics questioning variously whether
climate change is indeed occurring, whether Earth is in fact warming,
and whether, as a consequence, polar bears are at any greater risk of
depletion, extirpation, or extinction than they were, say, fifty years ago.
In 2007, a group of authors led by Marcus Dyck argued, among other
points, that spring air temperatures in the Hudson Bay region showed
no particular increase beyond the expected interannual variation, that
any changes in the timing of sea ice breakup in Hudson Bay were not
statistically significant, and that the contention that polar bears cannot

survive as a species without summer sea ice was dubious. Anyway, if the summers did become especially tough, polar bears could adapt by adopting a hibernative state, like their black and brown bear relatives.

As for observed changes in polar bears' health, they could just as easily be ascribed to other causes. Repeated darting and examination by scientists were surely bound to stress polar bears and quite possibly affect their physical condition—to say nothing of being constantly gawped at by tourists in vehicles that lumber past them on the tundra, or being captured and placed in the Polar Bear Jail every time they wander into Churchill.

But Dyck and his coauthors had not looked at temperatures over the same time period as Stirling and colleagues in their earlier assessment. Instead, they had, for no discernible reason, taken an apparently arbitrary starting point of 1932, which happened to coincide with a brief warm spell, rather than focus on the more recent time period when the biologists had recorded increasing temperatures in the air and decreasing conditions in the bears. Blaming those decreasing conditions on scientists and Tundra Buggy tourists contradicted all the available studies that had investigated those very impacts and also ignored the clear correlation that Stirling and others had shown between bears' physical state and the amount of time they were able to stay on the sea ice. As for the suggestion that polar bears could just learn to hibernate, like their relatives: unfortunately, an inconvenience of evolution is that such physical adaptations tend not to take place overnight, or even within the number of generations that stand to be affected by a rapidly changing climate. Besides, black and brown bears don't truly hibernate, either; as Stirling, Derocher, and others pointed out, "Large mammals are prevented from being true hibernators by the tremendous energetic costs associated with arousal from a low metabolic state"—a cost that would be even greater for the largest carnivore in the coldest environment. Which is why, they continued, the largest hibernating mammals are, in fact, marmots.

Little wonder that Derocher was quoted in the *Anchorage Daily News* as doubting that most of the article's authors "had ever seen a polar bear."* (Despite its clear shortcomings, however, the article was quoted often by then-governor of Alaska Sarah Palin in opposition to the notion of granting polar bears greater federal protection.)

Still, small but vocal groups of contrarians remained. One particularly persistent assertion was that, far from showing signs of decline or endangerment, the global polar bear population has grown over the past several decades—the factor most frequently expressed was quadrupled. The contention's origin was not in the peer-reviewed literature, but in the conservative blogosphere it became received wisdom. Bjorn Lomborg, a serial skeptic on environmental matters, wrote that there were "probably 5,000" polar bears in the 1960s, citing a *New York Times* article that read, in part: "Other experts see a healthier population. They note that there are more than 20,000 polar bears roaming the Arctic, compared to as few as 5,000 40 years ago." When Peter Dykstra, then executive producer for science, tech, and weather for CNN, asked the author of the *Times* article for the source of his figures, the latter said that he could not remember but that he understood the numbers to be "widely accepted." Certainly, Dykstra found, in a review for the Society of Environmental Journalists, that they were widely repeated, albeit with variations:

In a May 20 *Los Angeles Times* opinion piece, Jonah Goldberg took a whack at what he sees as quasi-religious overtones to conservation. Part

---

* Whether or not that was true, critics took note of the fact that two of the paper's co-authors were Willie Soon and Sallie Baliunas of the Harvard-Smithsonian Center for Astrophysics, two perennial advocates of the position that any global warming that may be occurring is within the boundaries of natural variation and likely primarily caused by changes in solar output. When *Climate Research* published a paper they coauthored, which argued that the twentieth century showed no signs of being atypically warm, half the journal's editorial board resigned in protest at what they felt was an insufficiently rigorous peer review of a paper that did not merit publication.

of his backup? "Never mind that polar bears are in fact thriving—their numbers have quadrupled in the last 50 years." . . . James Taylor of the Heartland Institute cited a London *Daily Telegraph* article that "confirmed the ongoing polar bear population explosion" in a September 11, 2007, blog . . . Taylor adds a new number into the mix from a March 26, 2008, posting at the Heartland site: "The global polar bear population has doubled since 1970, despite legal polar bear hunting" . . . From James Delingpole, a *Times* of London blogger, similar numbers, but different dates. And no source: "In 1950, let us not forget, there were about 5,000 polar bears. Now there are 25,000."

A clue to the likely origin of the much-repeated 5,000 could be found in a URL that Lomberg e-mailed to Dykstra, which linked to a paper tabled by Soviet scientist Savva Uspenski at the First Scientific Meeting on the Polar Bear in Fairbanks, Alaska, in 1965. Extrapolating from his own studies of denning sites, Uspenski proposed a global figure of 5,000 to 8,000. But, to repeat some text from two chapters previously, his view was by no means the consensus:

> Five thousand, said some. More than 10,000, reckoned Canada's Richard Harington. Between 10,000 and 19,000, according to official U.S. estimates. As high as 25,000, according to others.

Thor Larsen, who was present at both that first Fairbanks meeting and the subsequent inaugural meeting of the Polar Bear Specialist Group, told Dykstra, "Most data on numbers from the late 1960s and early 1970s were indeed anecdotal, simply because proper research was lacking. As far as I can remember, we did stick to a worldwide 'guesstimate' of 20–25,000 bears in these years." Ian Stirling ventured that any estimate of 5,000 bears "was almost certainly much too low." And Steven Amstrup commented that, at the time of the 1965 meeting, "people were just beginning to figure out how we might study animals scattered over the whole Arctic in difficult logistical situations. Some estimated that world population might have been as small as 5,000 bears, but this was nothing more than a WAG."

WAG, Dykstra points out, is scientific jargon for "wild-ass guess."

Since those early days, Amstrup argued, "the scientific ability to estimate the sizes of polar bear populations has increased dramatically." Even so, many uncertainties remain. While the Polar Bear Specialist Group agrees that the species' worldwide population is indeed somewhere within the range of 20,000 to 25,000, there are no reliable estimates at all for the Kara Sea region or the Arctic Ocean basin. Of the nineteen recognized subpopulations, the PBSG was unable, as of its 2009 meeting, to determine status or trends for seven. For those where the data are sufficiently robust, however, the picture is not bright. Three of the remaining twelve are reckoned to be stable, and one, in the Canadian Archipelago, even increasing slightly—albeit from much-reduced numbers as a result of previous overhunting. But fully nine are considered to be in decline, among them western Hudson Bay, the polar bears of Churchill, where additional years of study and number crunching have allowed researchers to calculate with some confidence that these most iconic of all polar bears have decreased in number from approximately 1,200 in 1987 to a little over 900 by 2004.

Some Inuit hunters, noting increases in polar bear sightings near settlements during the time when the ice has melted and the bay is open water, have argued that polar bear numbers are growing, not diminishing. Researchers, in contrast, remain equally convinced by data showing a lack of increase at best, and future or even current declines at worst. Those declines, they fear, will be made only worse by increases in quotas for subsistence hunts, increases made on the basis of the more frequent sightings, sightings that scientists assert are the result of hungrier, more desperate bears coming inshore in search of food.

Such encounters are not restricted to Canada. In northern Alaska, too, polar bears have begun to encroach more on human settlements, sometimes out of apparent desperation, sometimes seemingly because sheer exhaustion robs them of their normal caution and determination to avoid potential danger.

Alaska journalist Charles Wohlforth relays one particularly evocative instance of the latter, as told to him by Barrow biologist Craig George. Summer was reaching its end, and the sea ice of the southern Beaufort Sea had retreated fully 200 miles, by George's reckoning, from the coast. And yet, as if from nowhere, a mother and two cubs swam ashore on the beach near the town. Twice residents used cracker shells to drive the bears back into the ocean; twice they returned. Continued George, in Wohlforth's telling:

> Dodging cracker shells, she came ashore a third time and walked right through our crowd of dissuaders, crossed the beach road, and lay down with her cubs, barely 100 yards from the beach and only 10 yards from the road. She seemed to say, 'Shoot me if you must, but I ain't moving. If I go back to sea, I'm dead anyway . . .' So there she lay with her cubs for two days, barely moving a muscle . . . and finally, after two days of comatose rest, she slowly got up and ambled to the coast with her cubs to spend the rest of the fall on the tundra. We never saw her again.

In the waters north of Alaska and Siberia, the Beaufort Gyre has long circled endlessly, a closed loop that gathers ice floes into its embrace and pushes them into and onto each other, creating pressure ridges that rise from the surface of the floes and enormous islands of ice that grow ever thicker, the entire assemblage the perfect combination of stability and volatility, the ice thick enough to provide platforms for seals and bears but active enough to provide multiple openings where seals may breathe and bears may feed.

Since the beginning of the twenty-first century, however, the Beaufort Gyre has been weakening, and warming coastal waters have eaten away at the ice it contains. The thick, stable floes that used to reach all the way to the coast now retreat from it, changing the nature of northern Siberia and Alaska perhaps forever and bringing into question their future as polar bear residences.

"Places like northern Alaska and the northern coast of Russia, the sea ice has historically been close to the shoreline in summer; now it's way offshore and by mid-century it will probably be totally absent," says Steven Amstrup. "If the bears stay with the ice like they have been doing, and the ice retreats farther and farther to the north, is there any sense in them coming back for just a few months in the wintertime when the ice refreezes? Would you migrate a couple of thousand miles for just a short stay and return?"

Yet therein, especially for pregnant females, lies a dilemma. Those that choose to continue denning on the sea ice, as more than half of Beaufort Sea females traditionally have done, will do so on a platform that is thinner, more chaotic, and less reliable than before, that may transport them even farther than previously while they rest in their dens, that could conceivably break apart beneath them. It seems that, as the ice thins and retreats, a greater number of bears than before are now making dens ashore, but this, too, poses its risks. A growing distance between shoreline and ice edge exposes newly emergent cubs to the considerable dangers of being forced to take a plunge before reaching the sanctuary of the sea ice; and while their mother is likely to carry them on her back, her stamina is not limitless. Polar bears are excellent swimmers, it is true, but in swimming too great a distance without respite they risk exhaustion, as in the case of the family described by Craig George, or worse.

Between 1987 and 2003, biologists Charles Monnett and Jeffrey Gleason saw a total of 351 polar bears during aerial surveys off Alaska's Beaufort Sea coast, 12 of which were swimming at the time the aircraft passed over them. In 2004, they saw 55 bears, 51 of them alive, and 10 of those living bears in the water. The 4 dead bears were all floating in the water, and while it cannot be certain that they drowned, it can reasonably be surmised that they did. It is, suggest Monnett and Gleason, a fate that is likely to befall an ever greater number of polar bears as the ice retreats and disappears.

Our story began in what seemed on the surface the most fragile of environments, and it is where we shall end. A layer of snow covering a hole in a drift scarcely seemed adequate protection for a mother to raise her cubs, and yet, in the dark and the warmth, that is what she does. It is, in fact, in many ways the most peaceful and secure experience of a polar bear's life. But that security, too, may no longer be a guarantee.

As temperatures climb in western Hudson Bay, so too does the likelihood of fires in the areas of the Wapusk National Park where forest yields to tundra and, in the other direction, tundra vegetation is slowly subsumed by a forest of spruce and larch. It is here, in the shade and relative cool, that the polar bears of Churchill spend their summer months, curled up in dens dug into the permafrost, dens in which expectant mothers remain in wintertime, digging outward into the snowbanks that cover them up with the onset of fall. But fires caused by lightning strikes sometimes damage those earthen shelters, causing their collapse, and while polar bears returning from the sea ice will investigate dens in burned areas, they rarely decide to remain there, forcing them to keep searching or, possibly, to dig new ones instead of being able to take advantage of those that have existed for generations. It is yet another expenditure of energy that a pregnant female in particular can ill afford to undergo, but one that may prove increasingly difficult to avoid.

Far to Churchill's northwest, researcher Doug Irish was traveling along the coast of the Yukon in June 1989 when, sticking out of a snowdrift, he saw the head of a dead polar bear. Digging into the snow, he realized that he had stumbled across a den that had collapsed; buried in what remained of the dwelling were two small cubs. Given the extreme care and exactitude with which polar bear mothers choose den locations, such collapses are likely exceedingly rare; but in a world where snow is more prone to melt, or heavy rains take the place of some snowfall, they may become less so.

Yet even the most ideal den location, even the most perfect winter conditions, will be to no avail if, as the ice retreats, as seals diminish and become harder to find, as polar bears become hungrier and ever more desperate, a den is no longer a sanctuary but a target.

In January 2004, researchers flying over denning habitat along the Beaufort Sea coast came across a den that had been broken open. From the den's opening, a trail of blood led to the carcass of a bear. Landing to investigate further, they documented a scene without known precedent. Arterial blood was sprayed along the back wall of the den, into which a great deal of snow and ice rubble had been pushed. Two hundred feet away, the carcass was of a female, the den's occupant. Surrounded by large paw prints, it had been partly devoured.

Examination of a single set of paw tracks shone light in the shadows and completed the picture. An adult male—wandering, meandering, directionless—had suddenly stopped and made straight for the den. Using his massive forepaws, he had smashed through the roof as if it were the lair of a ringed seal. When the roof caved in, the snow buried two young cubs, but the male was after larger prey. His massive limbs held the mother down as he bit at her head and neck, severing her artery and penetrating her skull. Then he dragged her into the open and began to eat her.

Cannibalism in polar bears is not without precedent. It is to avoid this prospect that females keep cubs close by and build their dens far away from areas where male bears are likely to roam. But such incidents as do occur are almost always the result of accidental encounters; when females are killed and sometimes eaten, it is likely in defense of their young. Never before had any of the researchers involved seen anything quite like this, never before had there been any record of a male actively breaking into a den and killing a female.

Three months later, the same researchers followed the tracks of a female which, having lain undisturbed during her denning period, had emerged with one cub in tow. The tracks led to a pressure ridge, where she had lain down in the snow and nursed her cub. It would be the last meal she offered, and the last the cub consumed. Just beyond the ridge, she lay dead, overwhelmed, killed, and eaten by a much larger bear. The tracks of the fleeing cub continued for a short while but then were lost.

And just three days after that discovery, the researchers made another: an adult male feasting on a yearling, which had been stalked and killed while it lay in a small pit in the snow. A rash of paw prints in the vicinity may have belonged to the dead cub's mother and siblings; but of the bears themselves there was no sign.

Many of the living bears the researchers found that year were in relatively poor condition; while offering the usual caveats of scientific uncertainty, they did not hesitate to suggest a relationship between the existence of thin bears and cannibalized ones. In twenty-four years of fieldwork in the southern Beaufort Sea, they wrote, never before had they come across signs of polar bears actively stalking and killing other polar bears.

Previously unseen, also, was the tragic tableau witnessed on the tundra of Hudson Bay, just outside Churchill, on November 20, 2009. The fall season had been warm and long; the shores were densely packed with hungry bears awaiting the sea ice that had yet to form. The visitors that day did not see the attack itself, did not document the moment when the male had attacked and killed the cub, but observed the aftermath. A struggle of sorts was evident as they approached; by

the time they arrived on the scene, the cub was dead, the distraught
mother alternately charging and then circling its killer. Perhaps attracted by the scent of death, other bears arrived; the male, which had
begun to consume his quarry among some willows, moved the carcass
to the coast to finish his meal in relative peace. The mother, as if still
not able to fathom what had taken place, wandered the area looking
frantically for her offspring. Where the male had begun to feast, she
found her cub's pelt; picking it up in her mouth, she carried it away,
swinging her head from side to side in obvious distress. Charging the
other bears to keep them away, she kept the pelt firmly in her mouth
until, finally, she placed it gently in the snow among the bushes, protected from the wind.

In 2007, in response to demands from environmentalists that it protect polar bears under the federal Endangered Species Act, the Bush
administration directed the USGS to provide its best assessment of the
species' status and future. The response was not encouraging.

Based on the best available estimates of climate trends and the likely
response of sea ice to those trends, USGS scientists concluded that, by
the middle of the twenty-first century, two-thirds of the world's polar bear population may have disappeared. It is possible, they wrote,
that polar bear populations can survive in the Canadian Archipelago
through the end of the century; if they do so, they could be the species' last survivors, and they would be much reduced in number. Elsewhere, polar bears would likely cease to exist within seventy-five years,
and in some parts of their range — such as Hudson Bay and the southern Beaufort Sea — they could be gone within forty-five.

"I'm sure we'll still have polar bears around by the middle of the
century and probably by the end of the century," says Steven Amstrup, the study's lead author, "but they'll be limited to portions of
the Canadian High Arctic and adjacent Greenland, where the sea ice

remains for longer, so they'll still have enough time to forage on marine mammals . . . [and] can survive through those periods when the ice is absent."

In a few select places, such as perhaps Wrangel Island, they may be able to persist by feeding on walruses that have themselves been forced ashore; but for the vast majority, living on land will not be an option. The brown bears that live in the regions where any newly terrestrial polar bears would be forced to try and eke out a living are the smallest brown bears in the world, and they are sparsely distributed, because the environment can support only grizzlies that are relatively tiny and few in number.

"It seems highly unlikely, then, that you could take the largest bears in the world and plunk them down on land in a habitat that only supports a very spare population of some of the smallest bears in the world," observes Amstrup. "You wouldn't do that and expect the polar bears to be able to survive. One thing we know is, polar bears don't go running around trying to figure out how to catch different kinds of food. Nowhere that we're aware of have they been successful in garnering much energy from anything except marine mammals that they catch from the sea ice."

Polar bears, in other words, may have descended from grizzlies, but there is no turning back to be like them again. They took a fork in the evolutionary road, and the path they have followed leads in only one direction. Polar bears have evolved to exploit a particular environment, a specific niche. In so doing, they have become a supremely successful predator, but while they may be the dominant predators in their chosen realm, without it they are doomed.

Polar bears are creatures of the sea ice. If it disappears, so will they.

The warmth of the freshly slain seal enveloped his face; its scent flooded his nostrils. At long last, the wait was over. The sea ice had returned, and with it a chance to satiate his hunger. He tore chunks of blubber from his victim, swallowing them ravenously as if he had no time to chew, so desperate was he to fill the void inside him.

The wait had been longer this year, the fast more demanding. It had eaten at him, increased his yearning for the hunt, a yearning that burned inside him still even as he devoured his first kill of the season. In others it had created desperation; one young male, anxious not to have to travel any farther after months without food, had even attempted to approach the breathing hole he had so studiously staked out. Such an act of impudence, and one he had punished, driving the intruder away until, defeated, the thin youngster had lain down in the snow and seemingly surrendered to the inevitable. As he swallowed, he looked over his shoulder to check that the stranger was not attempting to snatch his meal, but the interloper was not moving, showed no sign of stirring at all in fact, but remained where he lay, not offering any resistance as the drifting snow slowly covered him up.

The meal was finished. Only the remnants of a carcass remained. Still he was hungry. He closed his eyes and lifted his nose, sampling the scents that wafted through the air. He opened his eyes again and scanned the horizon. It was flat, and still. He sniffed the air some more and then set off on his journey, across the ice.

Future

The Nares Strait is a narrow sliver of a passage, an undersized intermediary sliding between the closest points of Greenland and northeast Canada. The multichromatic mountaintops of Ellesmere Island glisten to the west, a giant's fingertips away from the slightly less imposing cliffs of Greenland to the east.

On a map it is barely noticeable, a cigarette paper's width lost in the morass of channels and islands of the Canadian High Arctic. For much of its recorded history, it has been a navigable passageway in name only; in winter, it is thick with sea ice that in summer melts and breaks apart one section at a time, granting only limited and temporary access.

The first section, toward the strait's southern end, normally breaks apart in late June, but its fracture provides no guarantee of a clear pathway. The ice takes weeks to fully break up and drift south, and it does so in the form of large floes and ice islands that are a navigational peril. By the time something approaching a negotiable route to the north opens up, a second restraining ice bridge—at the strait's very northern limit, on the boundary of the Robeson Channel and the Lincoln Sea—starts to fracture, flooding Nares Strait with old, multiyear sea ice from the Arctic Ocean.

In terms of oceanography and climate, that makes the Nares Strait, for all its apparent anonymity, of extreme importance, because it is one of only two outlets through which the Arctic Ocean can expel ice. But it also makes further passage precarious and ultimately—as temperatures drop with the onset of autumn and the ice fuses into an impenetrable barrier—once more impossible.

Prior to 1948, only five vessels had ever been recorded as traveling even as far north as Kane Basin, a slight bulge that marks the strait's approximate midway point. The ships that had pushed north of that point in the subsequent half-century were also relatively few in number, primarily powerful icebreakers that could grind through the gnarled floes. But now, at the end of June 2009, a small green ship—an icebreaker, yes, but on a far smaller scale than the behemoths that preceded it—sails defiantly and without interruption. On each flank of the ship's hull a rainbow rises from the waterline toward the bow, culminating in a white dove. Above the doves, in white lettering on either side, the ship's name.

It is the *Arctic Sunrise*.

It had left Amsterdam two and a half weeks previously, and when its latest expedition had been conceived, there was no certainty it would be able to reach its destination, no guarantee that the *Sunrise* would be able to penetrate the southernmost ice barrier, or that, were it successful in that goal, it could successfully navigate the ice that would surely be scattered about the strait. But during the winter, ice had never fully consolidated, and by the time the *Sunrise* arrived, nary an ice floe was to be seen.

So the little ship sailed north, at no stage impeded by or even in sight of ice, until, 300 miles later, it had finally traveled as far north as it could. It had traversed the Nares Strait from south to north, the first ship ever to do so in June, but the pride in priority was countered on board by awareness of the possible ramifications. One swallow does not a summer make, but it was the second time in three years that winter ice in the Nares Strait had failed to consolidate fully. On the previous occasion, in 2007, a constant torrent of floes from the Lincoln Sea had flushed through from north to south all summer, but on this occasion, the ice bridge in Robeson Channel had held fast, and it was here, finally, that the *Arctic Sunrise* found its northward progress halted.

Ahead stretched nothing but the sea ice of the Arctic Ocean, 450 miles separating the *Sunrise* from the North Pole. There was need for caution: once the ice began to fracture and flood the strait, the *Sunrise* would have to flee or, at least, take shelter. A helicopter survey showed that, for now at least, the bridge was holding steady; from the air, too, there were abundant signs of the productivity of the ice edge, pathways in the snow where seals had slid into the water and, crisscrossing the ice like the feverish scribbling of a demented mapmaker, pathway after pathway of polar bear tracks, emerging from the distance, converging on the edge of the fast ice and patrolling its length, heading to breathing holes, crossing over pressure ridges, circling back and around, and eventually disappearing again.

And then, the crew of the *Sunrise* scarcely having had time to acknowledge their surroundings and finish breakfast, one of the bears appeared, carefully walking in the same prints that either it or another had left earlier and heading curiously toward the strange interloper into its world. For those who had been on board the ship in similar surroundings before, it was an omen: the first polar bear of the expedition, appearing on the first morning at the destination. Being the first, its every move was met with a cacophony of beeps and clicks as two dozen camera shutters fired off in rapid and ongoing succession.

The bear seemed fat and healthy, confirmation that the ship had entered productive hunting grounds, but for a good half-hour it appeared interested less in hunting than in examining the strange green iceberg that had appeared as if by magic. It sniffed the new smells the iceberg had brought with it; it examined the objects that strolled around on top and emitted peculiar noises. Occasionally, it wandered off to examine the edge of the ice, peering over the edge as if at its own reflection in the water and then looking back over its shoulder and baring its teeth as if the presence of the green iceberg, no longer a novelty, was now a source of confusion or displeasure.

For sixteen days the ship remained near the northern end of the strait, stationed alongside a glacier that the team of onboard scientists studied daily and intently, measuring movement and melt and the temperature of the water that lapped at its head until, eventually, the ice bridge broke. The Arctic Ocean displaced the *Arctic Sunrise*, its floes assuming their rightful place in the Nares Strait, their looming presence enough to prompt the ship's departure.

It would be a staged retreat; the ship would stay ahead of the ice and then, as if pressing itself in a doorway to avoid the passage of slow-moving Pamplona bulls, tuck itself into Kane Basin a while until it could stay no longer. And it was there, on a bright, sunlit Arctic night, that scientists and crew set off on a short boat ride to investigate an iceberg of exceptional beauty, nicknamed by those on board the "doughnut berg" for the almost perfect arch that it formed. Unlike many other icebergs, this was likely not a function of years of weathering but a feature with which it was created, the result of a channel in the glacier from which it calved.

The boat returned to the ship to pick up a photographer to examine the berg more closely, but as it neared its destination, the cry went out from above that a polar bear was approaching, at full speed. Vulnerable in their small craft, the boat crew fought their way through chunks of ice to reach the sanctuary of the pilot door and clamber onto the *Sunrise*; their safety assured, the mood of all on board shifted from anxiety to the relaxed confidence afforded by the protection of an icebreaker.

The bear approached, wading knee-deep through a melt pool in the ice, striding confidently toward the bow. It sniffed the air, looked up at those looking down, and for an instant appeared to crouch, as if measuring the distance and preparing to spring upward. Even from the safety of the bridge wing, thirty feet above its head, the bear's seemingly predatory focus unnerved. The bear completed its calculus and, having satisfied itself that any leap would fall short of its goal and be energy unnecessarily expended, snapped back to reality until, startled by the knock of a tripod leg against the steel hull, it whipped round and splashed into the water.

Hauling itself onto a slightly more distant piece of ice, it rocked back onto its haunches and, its back now ramrod-straight, pointed its nose straight up toward the bright blue sky. It looked as if it could be about to howl at the midnight sun or was perhaps adopting a pose of yogic meditation instead of, as was in fact the case, gaining maximum elevation to assess the scents that wafted from the ship. It dropped its forepaws to the ice once more and, without warning, rolled onto its back in the snow, shifting its rump back and forth as its massive paws waved harmlessly in the air. It was likely cooling off after its exertions, but to those on board the *Arctic Sunrise*, what had just minutes ago been a cause of concern and rapid disembarkation now appeared as threatening as an oversized family pet.

Returning to its feet, it took a second to regain its composure before wandering off across the ice. And then, in an instant, it reverted to its previous mode. A dark lump in the distance revealed itself through binoculars to be a seal, seemingly asleep on the floe; the bear flattened itself against the ice and crawled stealthily and circuitously toward its prey. Whether alerted to the bear's presence or simply determining it was time to move, the seal slipped swiftly into the water and away. The bear walked up to the edge of the ice floe, sniffed the air, and, with the ursine equivalent of a mildly disappointed shrug of the shoulders, continued its wandering, until it was out of sight.

There was a rumble as the Arctic stillness was broken by the sound of the ship's engines starting.

The ice was encroaching.

It was time to leave.

The ship turned around and began to steam south, the Nares Strait and its polar bears slowly receding into the distance.

Notes

page **BECOMING**

12 "a superabundance": Ian Stirling, *Polar Bears*, p. 84.

15 "Historically . . . to den on": Steven Amstrup, conversation with the author, March 9, 2009.

16 "interconnecting tunnels": Thomas S. Smith, comments on draft manuscript.
    "may not be a ventilation shaft": Ibid.

17 "severity"; "tall, jagged": Nikita Ovsyanikov, *Polar Bears*, p. 96.

18 "hissed so loudly": Ibid., p. 99.

19 "I think . . . against the far wall": Richard Harington, conversation with the author, August 16, 2008.
    "I remember": Ibid.

20 "I was with Tam Eeolik": Ibid.
    "In the course of that . . . at home or not": Geoff York, conversation with the author and others, Tundra Buggy Lodge, October 31, 2008.
    "We shouted and hollered . . . like a turtle": Ibid.
    "She saw me . . . she took off": Ibid.

23 "A human could *never* . . . have no problem": Thomas S. Smith, comments on draft manuscript.

23   "During the months . . . you keep at it": Mike Spence, conversation
      with the author, October 26, 2008.

24   "I would not have been able": Thorsten Milse, *Little Polar Bears*, p. 59.

25   "rather inconsiderately"; "became aware"; "a bear's face": Hugh Miles
      and Mike Salisbury, *Kingdom of the Ice Bear*, p. 44.
      "was greeted"; "The female fussed"; "the female emerged": Ibid., p. 47.

26   "in the realm of the seal": Ibid., p. 48.
      "When the moment finally arrives": Milse, *Little Polar Bears*, p. 19.
      "Once she comes out . . . off she'll go": Mike Spence, conversation with
      the author, October 26, 2008.

27   "will spend mere minutes": Thomas S. Smith, comments on draft
      manuscript.
      "I'm convinced . . . off they go": Ibid.

30   "If there are seven documented instances": Stirling, *Polar Bears*, p. 137.

BEAR

35   "beyond the north wind"; "Illnesses cannot touch them . . . this exalted
      race": Kieran Mulvaney, *At the Ends of the Earth*, p. 1.

45   "gigantic": Steven C. Amstrup, "Polar Bear: *Ursus maritimus*," in *Wild
      Mammals of North America*, p. 591.

47   "If a grizzly . . . pull it back out": JoAnne Simerson, San Diego Zoo,
      conversation with the author, December 15, 2008.

48   "To test the idea . . . made by its breath!": Stirling, *Polar Bears*, p. 144.

49   "Typically . . . in the water": JoAnne Simerson, conversation with the
      author, December 15, 2008.
      "They're like snowshoes": Ibid.

50   "Warm surfaces . . . inside the pelt": David Lavigne, International Fund
      for Animal Welfare, conversation with the author, December 11, 2008.

51   "a miniature light pipe": Richard C. Davids, *Lords of the Arctic*, p. 25.
      "We discovered . . . 'Black Polar Bears'": David Lavigne, conversation
      with the author, December 11, 2008.

52   "Now the ultraviolet . . . *absorbed* by the hair": Ibid.

53   "of a monstrous bigness": Richard Hakluyt, *Voyages in Search of the North-
      West Passage*, retrieved online from http://ebooks.adelaide.edu.au/h/
      hakluyt/northwest/chapter7.html.

ICE

57 "many of our company": Robert McGhee, *The Last Imaginary Place* (Chicago: University of Chicago Press, 2005) p. 158.

58 "terrifyingly like": Mariana Gosnell, *Ice*, p. 181.

"most fearfull both to see and heare": Gerrit de Veer, *A True and Perfect Description of Three Voyages (1609)* (Delmar, NY: Scholars' Facsimiles and Reprints, 1993) p. 101.

"Driven by the power of winds . . . against such a force": McGhee, *The Last Imaginary Place*, p. 137.

59 "It's like a wonderful time-lapse version . . . It's pretty intense": Brendan P. Kelly, conversation with the author, November 13, 2008.

"For me": Eric Larsen, conversation with the author, November 11, 2008.

"There are pieces of ice . . . It's always changing": Ibid.

60 "We had a bear . . . in less than a minute": Ibid.

67 "like a lunar landscape": Brendan Kelly, conversation with the author, November 13, 2008.

70 "I've seen bears . . . they don't like being near them": Ibid.

"mostly for their sing-song linkage"

71 "a circus"; "They whistle": Ibid.

Natalie Angier, "Who Is the Walrus?" *New York Times*, May 20, 2008.

"The call came . . . it was wild": Brendan Kelly, conversation with the author, November 13, 2008.

72 "are always in battle": Erik W. Born, *The Walrus in Greenland*, p. 4.

73 "A polar bear had swum too close": Erik W. Born, *The White Bears of Greenland*, p. 31.

"were on a ship": Brendan Kelly, conversation with the author, November 13, 2008.

74 "If you go . . . right out of the water": Ibid.

76 "When the first molecule-thick . . . nose out to breathe": Ibid.

77 "If you put your head . . . a little chapel": Ibid.

LIFE

79 "virtually immortal": Stirling, *Polar Bears*, p. 139.

82 "When such groups . . . madly in the air": Brendan Kelly, comments on draft manuscript.

82   "that only the tip . . . after its prey": Stirling, *Polar Bears*, p. 116.

83   "I was amazed . . . concious memory": Ibid., p. 117.

84   "A lair might be a meter": Brendan Kelly, conversation with the author, November 13, 2008.

85   "Before the bear . . . of the blow": Born, *The White Bears of Greenland*, p. 43.

86   "What happened next": Davids, *Lords of the Arctic*, pp. 67–68.

87   "Yesterday I watched": Charles Feazel, *White Bear*, pp. 2–3.
     "I often find": Brendan Kelly, comments on draft manuscript.

89   "seen them back up": Davids, *Lords of the Arctic*, p. 3.

90   "The value of this alternate food": Amstrup, "Polar Bear: *Ursus maritimus*," p. 592.
     "it is essentially": Geoff York, comments on draft manuscript.

91   "Imagine . . . a treadmill": Ibid.
     "Rear rudders of U.S. submarines": Andrew Chang, "Polar Bear Attacks U.S. Submarine," ABC News, May 30, 2003.
     "I can't even imagine that . . . Unreal": Eric Larsen, conversation with the author, November 11, 2008.

96   "absolutely desolate"; "we saw fox tracks"; "never saw a fox": Bruce Weber, "Riding Bombardier snowmobiles, he was the first to reach the North Pole," *Globe and Mail*, October 22, 2008, p. R5.

99   "far from sight": Stirling, *Polar Bears*, p. 67.

100 "Polar bears used to be considered": Davids, *Lords of the Arctic*, p. 68.
     "a solitary ice wanderer": Ovsyanikov, *Polar Bears*, p. 90.

101 "It isn't that polar bears": Geoff York, comments on draft manuscript.
     "In fact . . . simple tolerance": Ovsyanikov, *Polar Bears*, p. 81.

102 "only a few experienced animals": Ibid., p. 66.
     "jumped down the small cliff": Ibid., p. 65.
     "In the ensuing panic . . . to defend it": Ibid., p. 66.

103 "It is quite possible . . . getting a meal": Ibid.

104 "I have seen": Geoff York, comments on draft manuscript.

ENCOUNTERS

109 "Bears, the hunters kept telling him": Davids, *Lords of the Arctic*, p. 74.

112 "If you track a bear . . . deeper and wider": Darren Keith et al., *Inuit Knowledge of Polar Bears*, p. 94.

"The heel will dig . . . the heel bone": Ibid.

"If you are . . . at some point": Ibid., p. 118.

113 "I have seen . . . when it came out": Ibid., p. 112.

"Before . . . and got stuck": Ibid., p. 113.

114 "Recently . . . the scientific literature": Richard Monastersky, "International Polar Year: The social pole?" *Nature* 457, no. 1078 (February 25, 2009).

115 "Arctic and Subarctic societies": Hugh Brody, *Living Arctic: Hunters of the Canadian North* (Seattle: University of Washington Press, 1987), p. 37.

"Europeans, with their agricultural heritage": Ibid., p. 43.

"Only the very strongest": Jeannette Mirsky, *To the Arctic!: The Story of Northern Exploration from Earliest Times* (Chicago: University of Chicago Press, 1970), p. 9.

117 "[B]ut our master": de Veer, *A True and Perfect Description of Three Voyages*, p. 169.

"We leavelled at her": Ibid., pp. 154–55.

118 "Her death": Ibid., p. 183.

119 "A great leane": Ibid., pp. 62–63.

120 "perceiving them": Ibid., p. 63.

"Our men are already dead": Ibid.

"making a great noyse": Ibid., p. 64.

121 "This fierce tyrant"; "This bear"; "The annals of the north": Anonymous, *The Mariner's Chronicle* (New Haven: George W. Gorton, 1835), p. 415.

123 "a race": Roald Amundsen, *My Life as an Explorer* (New York: Doubleday, Doran & Co., 1928), p. 86.

"I had always heard": Ibid., p. 88.

125 "If the bear is hunting": Stirling, *Polar Bears*, p. 117.

"I honestly think . . . well blended in": Eric Larsen, conversation with the author, November 11, 2008.

126 "When the stove . . . into the tent": Ibid.

127 "He showed no signs": Ovsyanikov, *Polar Bears*, p. 53.

"And then this giant . . . terribly frightened": Ibid.

"One time . . . crawled back out": Robert Buchanan, conversation with the author, December 22, 2008.

128 "there's just not that much": Ibid.

128 "The ones . . . deep doo-doo": Ibid.

"A typical encounter . . . so suddenly": Stephen Herrero, conversation with the author, October 27, 2009.

"thin"; "skinny": Stephen Herrero and Susan Fleck, "Injury to people inflicted by black, grizzly or polar bears," p. 31.

129 "Bears are curious . . . improving your safety": Stephen Herrero, conversation with the author, October 27, 2009.

"support the conclusion": Herrero and Fleck, "Injury to people inflicted by black, grizzly or polar bears," p. 31.

"People talk about . . . piss-poor job of it": Thomas S. Smith, conversation with the author, October 31, 2008.

"at least 251": Herrero and Fleck, "Injury to people inflicted by black, grizzly or polar bears," p. 31.

130 "Europeans took to . . . arctic journey": Barry Lopez, *Arctic Dreams*, p. 111.

"white bears of a monstrous bigness"; "being desirous"; "whereupon": Hakluyt, *Voyages in Search of the North-West Passage*.

"laid her paws"; "with signs": Lopez, *Arctic Dreams*, p. 112.

"though at first": William Scoresby, *An Account of the Arctic Regions*, vol. I (Newton Abbot, England: David & Charles Reprints, 1969), p. 522.

131 "he yielded": Ibid., p. 523.

132 "one of the finest": Davids, *Lords of the Arctic*, p. 101.

"The polar bear": Ibid.

133 "I am informed": Ibid.

"an international conference": Thor S. Larsen and Ian Stirling, *The Agreement on the Conservation of Polar Bears—Its History and Future*, p. 5.

135 "scientific knowledge"; "take such steps"; "conduct": Ibid.

"with a view": Ibid.

137 "The taking . . . be prohibited": International Agreement on the Conservation of Polar Bears and Their Habitat, Article I, Paragraph 1.

"take appropriate action": International Agreement on the Conservation of Polar Bears and Their Habitat, Article II.

138 "still no other": Stirling, *Polar Bears*, p. 190.

140 "The bigger": Jon Talon, conversation with the author, October 26, 2008.

141 "They all . . . these animals": Ibid.

"it's the closest": Geoff York, Tundra Buggy Lodge, October 31, 2008.

142 "Every bear . . . in fighting": Geoff York, Tundra Buggy Lodge, October 31, 2008.

143 "There are some . . . doing it again": Steven Amstrup, conversation with the author, March 9, 2009.

CHURCHILL

147 "The thing is . . . really careful": Doug Ross, conversation with the author, October 23, 2008.

"There were times . . . just in case": Ibid.

152 "In Winnipeg": Tony Bembridge, conversation with the author, October 26, 2008.

"My girlfriend . . . this door": Jon Talon, conversation with the author, October 26, 2008.

153 "Please leave a note": Lance Duncan, conversation with the author, October 25, 2008.

154 "once I cleaned . . . in that position": Ibid.

155 "Gone are those days . . . in a bear season": Mike Spence, conversation with the author, October 26, 2008.

156 "we just want": Shaun Bobier, conversation with the author, October 28, 2008.

"We had bears . . . the holding facility": Ibid.

157 "The dump . . . totally different": Ibid.

"In 2005": Ibid.

158 "Just prior . . . chase them through": Ibid.

159 "I talked . . . work since": Mike Spence, conversation with the author, October 26, 2008.

"The community's . . . respect that": Ibid.

"passing under the last streetlight"; "They said . . . the truck": Lance Duncan, conversation with the author, October 25, 2008.

162 "We would"; "They were skittish . . . wouldn't run away": Len Smith, conversation with the author, January 19, 2009.

163 "The first year . . . one basket anyway": Ibid.

"We had . . . people in": Ibid.

165 "So, now . . . damaged windshield": Don Walkoski, conversation with
     the author, October 29, 2008.
     "I sure": Len Smith, conversation with the author, January 19, 2009.
     "Welcome": Robert Buchanan, conversation with the author,
     October 26, 2008.
166 "I first started . . . dead on target": Ibid.
     "essentially a fraternity": Ibid.
167 "finance and educational"; "distributed mechanism": Ibid.
     "We're doing . . . be Arctic ambassadors": John Gunter, conversation
     with the author, October 26, 2008.
168 "The thing . . . just playing": Robert Buchanan, conversation with the
     author, October 26, 2008.
     "We use . . . incredibly smart": Ibid.

MELT
181 "Except we": Robert Buchanan, speaking with visitors aboard Buggy
     One, October 26, 2008.
     "What we're . . . about it": Ibid.
     "I want . . . polar bear": Ibid.
182 "The problem is"; "When . . . just can't": Ibid.
184 "as opaque": Spencer Weart, *The Discovery of Global Warming*, p. 3.
     "is a blanket": Ibid., p. 4.
190 "I find it incredible": "Arctic Sea Ice Down to Second-Lowest
     Extent; Likely Record-Low Volume." Press release, National Snow
     and Ice Data Center, October 2, 2008. http://nsidc.org/news/
     press/20081002_seaice_pressrelease.html.
191 "We are entering . . . in one go": Dave Walsh, "Postcard from the Ice
     Edge," Greenpeace Climate Rescue Weblog, September 20, 2009.
     http://weblog.greenpeace.org/climate/2009/09/postcard_from_the_
     ice_edge.html.
193 "In the Bering . . . disadvantaged": Brendan Kelly, conversation with
     the author, November 13, 2008.
195 "a substantial": Jonathan Verreault et al. "Chlorinated hydrocarbon
     contaminant and metabolites in polar bears (*Ursus maritimus*) from
     Alaska, Canada, East Greenland, and Svalbard: 1996–2002. *Science of
     the Total Environment* 351–352 (2005): 369–390.

197 "like science-fiction": Kieran Mulvaney, "Conservation as a Human
    Problem," in *Beyond the Bars: The Zoo Dilemma* (London: Thorsons, 1987),
    p. 157.
198 "that polar bears": Ian Stirling and Andrew Derocher, "Possible
    impacts of climate warming on polar bears," p. 242
    "fewer adult": Ibid.
199 "If climatic warming . . . their range": Ibid., p. 240.
    "From 1981"; "The trend": Ian Stirling, Nicholas Lunn, and John
    Iacozza, "Long-term trends in the population ecology of polar bears in
    western Hudson Bay in relation to climatic change," p. 294.
200 "Earth's climate"; "these climate changes": ACIA, *Impacts of a Warming
    Arctic*, p. 8.
    "reduction in sea ice": Ibid., p. 10.
    "about the possible": Press release, 11th meeting of PBSG,
    Copenhagen, Denmark, 1993. http://pbsg.npolar.no/en/meetings/
    press-releases/11-Copenhagen.html.
    "ecological change"; "greatest future": Press release, 13th meeting of
    PBSG in Nuuk, Greenland, 2001. http://pbsg.npolar.no/en/meetings/
    press-releases/13-Nuuk.html.
201 "As the distance": Andrew Derocher, Nicholas Lunn, and Ian Stirling,
    "Polar bears in a warming climate," p. 166.
202 "by the 2050s": Ibid.
    "it is unlikely": Ibid., p. 164.
203 "Large mammals": Ian Stirling, Andrew Derocher, William Gough,
    and Karyn Rode, "Response to Dyck et al. (2007) on polar bears and
    climate change in western Hudson Bay," p. 199.
204 "had ever": Tom Kizzia, "Funding and Review of Palin-Touted Study
    Criticized," *Anchorage Daily News*, January 27, 2008, p. A14.
    "probably 5,000"; "Other experts"; "widely accepted"; "In a May 20":
    Peter Dykstra, "Magic number: A sketchy 'fact' about polar bears keeps
    going . . . and going . . . and going," *SEJournal*, August 15, 2008. http://
    www.sej.org/publications/alaska-and-hawaii/magic-number-a-sketchy-
    fact-about-polar-bears-keeps-goingand-going-an.
205 "Five thousand": Ibid.
    "Most data"; "was almost certainly"; "people were just": Ibid.

206 "the scientific ability": Ibid.

207 "Dodging cracker shells": Charles Wohlforth, "On Thin Ice: Polar Bears in a Changing Arctic," in Kazlowski, *The Last Polar Bear*, p. 75.

208 "Places like . . . and return?": Steven Amstrup, conversation with the author, March 9, 2009.

212 "I'm sure . . . ice is absent": Ibid.

213 "It seems . . . from the sea ice": Ibid.

Annotated Bibliography

Some bibliographies are more useful than others. I am not a fan of listing pages upon pages of titles, some obscure and arcane, that do not provide the reader with any sense of which ones may be of greatest relevance. So what I have done here is provide an annotated guide to the books I found of particular use and interest during the course of this project, and which I suspect will be most likely to appeal to readers of this book.

Following that, I have also listed the key articles, reports, and scientific papers I consulted during my research, primarily the ones to which I referred most frequently and those whose findings I referenced directly in the text. I have kept these separate from the book listings because I recognize that most of the more technical papers will be of limited interest to many readers.

BOOKS

Anderson, Alun. *After the Ice: Life, Death, and Geopolitics in the New Arctic.* New York: Smithsonian Books, 2009.
  A journalistic look at the science and politics of climate change in the Arctic.

Archer, David, and Stefan Rahmstorf. *The Climate Crisis: An Introductory Guide to Climate Change.* Cambridge: Cambridge University Press, 2010.
Two highly respected climate researchers provide a concise, informed, and accessible primer on this most vital of topics.

Bieder, Robert E. *Bear.* London: Reaktion Books, 2005.
A slim, enjoyable guide to the natural and unnatural history of all bear species, extinct and extant.

Born, Erik W. *The Walrus in Greenland.* Nuuk, Greenland: Ministry of Environment and Natural Resources, 2005.
This well-illustrated guide crams an enormous amount of information on the polar bear's great foe into fewer than 80 pages.

Born, Erik W. *The White Bears of Greenland.* Nuuk, Greenland: Ministry of Environment and Natural Resources, 2008.
Hugely informative, packed with color photographs, with a focus on Greenland but a wealth of material relevant to the study of polar bears worldwide.

Brunner, Bernd. *Bears: A Brief History.* New Haven: Yale University Press, 2007.
Charming romp through ursine and ursine-human history.

Davids, Richard C. *Lords of the Arctic: A Journey Among the Polar Bears.* London: Sidgwick & Jackson, 1982.
Despite being relatively brief, in my mind one of the most informative popular polar bear books available, although of course now dated.

De La Lez, Mireille, and Fredrik Granath. *Vanishing World: The Endangered Arctic.* New York: Abrams, 2007.
Large-format volume with some explicatory text, but primarily devoted to some spectacular photographic imagery of the Arctic and its wildlife.

Domico, Terry. *Bears of the World.* New York: Facts on File, 1988.
Large format, well illustrated, and easily comprehensible text.

Eliasson, Kelsey. *Polar Bears of Churchill.* Churchill, Manitoba: Munck's Café, 2006.
Short, helpful background on the "Polar Bear Capital of the World" and its most famous denizens.

Ellis, Richard. *On Thin Ice: The Changing World of the Polar Bear.* New York: Alfred A. Knopf, 2009.
One of the most recent and comprehensive entries into the canon of polar bear literature. Very good resource.

Fast, Dennis, and Rebecca L. Grambo. *Wapusk: White Bear of the North.* Winnipeg, Manitoba: Heartland Associates, 2003.
Guide in words and photographs to the polar bears of Hudson Bay and beyond.

Feazel, Charles T. *White Bear: Encounters with the Master of the Arctic Ice.* New York: Henry Holt, 1990.
I'm a fan of this book; a nice guide to polar bears, their environment, and their interactions with humans, well written and with some strong firsthand accounts.

Gosnell, Mariana. *Ice: The Nature, the History, and the Uses of an Astonishing Substance.* New York: Alfred A. Knopf, 2005.
A gem; only partly about the sea ice that is the realm of the polar bear, but a splendidly written look at ice in its many and varied forms.

Guravich, Dan, and Downs Matthews. *Polar Bear.* San Francisco: Chronicle Books, 1993.
Basic, light on text, heavy on stunning photographs, introduction, and overview.

Henson, Robert. *The Rough Guide to Climate Change.* New York: Rough Guides, 2006.
An excellent introduction to, and explanation of, the subject that has so many concerned for polar bears' future.

Hopfner, Glenn. *Tales from the Tundra.* St. Rose, Manitoba: Glenn Hopfner, 2005.
A Tundra Buggy driver shares his experiences.

Kazlowski, Steven. *The Last Polar Bear: Facing the Truth of a Warming World.* Seattle: Braided River, 2008.
Terrific book. Large format, with sensational photographs. Atypically of this genre, it also includes some first-rate journalistic essays on polar bears.

Keith, Darren, et al. *Inuit Qaujimaningit Nanurnut: Inuit Knowledge of Polar Bears.* Gjoa Haven, Nunavut: Gjoa Haven Hunters' and Trappers' Organization/CCI Press, 2005.

Very nice compilation of interviews with Inuit hunters and trappers, with some real observational gems.

Koch, Thomas J. *The Year of the Polar Bear.* Indianapolis: Bobbs-Merrill, 1975. *Uses the travails of a polar bear over the course of a calendar year to describe the species' natural history, ecology, and interaction with humans.*

Lopez, Barry. *Arctic Dreams.* London: Picador, 1986. *Although less than 10 percent is devoted solely to polar bears, this book remains the literary standard by which Arctic nonfiction is judged.*

MacIver, Angus and Bernice. *Churchill on Hudson Bay: The Stories and History of Churchill, Manitoba as Told by Two of This Community's Pioneers.* Churchill, Manitoba: Churchill Ladies' Club, 2006. *Useful history of what has become the "Polar Bear Capital of the World."*

Mangelsen, Thomas D., and Fred Bruemmer. *Polar Dance: Born of the North Wind.* Omaha, NE: Images of Nature, 1997. *While this big and beautiful book is mostly a showcase for Mangelsen's first-rate photographs, Bruemmer's text is literate and engaging.*

Maslin, Mark. *Global Warming: A Very Short Introduction.* New York: Oxford University Press, 2004. *Slim and pocket-sized, but a comprehensive, clear, and highly valuable explanation of the science of climate change.*

Miles, Hugh, and Mike Salisbury. *Kingdom of the Ice Bear.* Austin: University of Texas Press, 1985. *Account of the filming of a BBC documentary series about the Arctic and its wildlife, particularly the carnivore that bestrides the sea ice realm.*

Milse, Thorsten. *Little Polar Bears.* Munich: Bucher, 2006. *Recipe for a successful book: 1. Include a small amount of text and a large number of photographs. 2. Ensure the photographs are exclusively of adorable polar bear cubs and their mothers. Absolutely delightful volume.*

Mulvaney, Kieran. *At the Ends of the Earth: A History of the Polar Regions.* Washington, DC: Island Press, 2001. *An environmental history of the Arctic and Antarctic.*

Ovsyanikov, Nikita. *Polar Bears: Living with the White Bear.* Stillwater, MN: Voyageur Press, 1996.
A classic. A firsthand account of studying the polar bears of Wrangel and Herald islands in the Russian Arctic.

Perry, Richard. *The World of the Polar Bear.* Seattle: University of Washington Press, 1966.
An early overview of what was then known of polar bears and their natural history — which was really not a great deal.

Rich, Tracey, and Andy Rouse. *Polar Bears.* Rickmansworth, UK: Evans Mitchell Books, 2006.
Short, well-illustrated monograph. A nice introduction.

Rosing, Norbert. *The World of the Polar Bear.* Buffalo, NY: Firefly Books, 2006.
Gorgeous. Rosing is an excellent photographer, and his images of polar bears and the Arctic in this coffee-table book are stunning. His text is also informative and digestible.

Sale, Richard. *The Arctic: The Complete Story.* London: Frances Lincoln, 2008.
"Complete" certainly describes this immense tome, large in format and extensive in content. A great deal of text and a great many photographs and illustrations make this the ultimate guide to the northern polar realms.

Sale, Richard. *A Complete Guide to Arctic Wildlife.* Buffalo, NY: Firefly Books, 2006.
A thick, comprehensive field guide to Arctic fauna.

Stirling, Ian, ed. *Bears: A Complete Guide to Every Species.* London: HarperCollins, 1993.
Several expert contributors lend authority to this large-format introductory guide.

Stirling, Ian. *Polar Bears.* Ann Arbor: University of Michigan Press, 1988.
More than twenty years after publication, this remains the definitive introduction, by the recognized authority on the species.

Streever, Bill. *Cold: Adventures in the World's Frozen Places.* New York: Little, Brown & Company, 2009.
Polar bears are but supporting players in the pages of this book, but it is a very well-written and very readable account of Alaska, the Arctic, ice, snow, and cold.

Thomas, David N. *Frozen Oceans: The Floating World of Pack Ice.* Buffalo, NY: Firefly Books, 2004.
It's all about the sea ice. Without it, there are no polar bears, no ringed seals, and a very different Arctic environment. This illustrated volume makes the subject surprisingly accessible and entertaining, even as it remains authoritative and scholarly.

Wadhams, Peter. *Ice in the Ocean.* London: Gordon and Breach, 2000.
A more complete, but slightly more challenging, work than the Thomas book.

Ward, Kennan. *Journeys with the Ice Bear.* Minocqua, WI: NorthWord Press, 1996.
Ward's excellent photographs are supplemented by an informative, engaging, and enthusiastic text.

Weart, Spencer R. *The Discovery of Global Warming.* Cambridge, MA: Harvard University Press, 2003.
A very, very good guide to the history of the science behind climate change.

OTHER PUBLICATIONS

ACIA. *Impacts of a Warming Arctic: Arctic Climate Impacts Assessment.* Cambridge University Press, 2004. 140 pages.

Amstrup, Steven C. "Polar Bear: *Ursus maritimus.*" In *Wild Mammals of North America: Biology, Management, and Conservation,* edited by G. A. Feldhammer, B. C. Thompson, and J. A. Chapman, 587–610. Baltimore: Johns Hopkins University Press, 2003.

Amstrup, Steven C., Bruce G. Marcot, and David C. Douglas. *Forecasting the Range-Wide Status of Polar Bears at Selected Times During the 21st Century.* Reston, VA: U.S. Geological Survey, 2007. 126 pages.

Amstrup, Steven C., Ian Stirling, Tom S. Smith, Craig Perham, and Gregory Thiemann. "Recent observations of intraspecific predation and cannibalism among polar bears in the southern Beaufort Sea." *Polar Biology* 29 (2006): 997–1002.

Bankes, Nigel. "Climate Change and the Regime for the Conservation of Polar Bears." In *Climate Governance in the Arctic,* edited by Timo

Koivurova, E. Carina H. Keskitalo, and Nigel Bankes, 351–82. Dordrecht, Netherlands: Springer, 2009.

Clarkson, Peter L., and Doug Irish. "Den collapse kills female polar bear and two newborn cubs." *Arctic* 44, no. 1 (1991): 83–84.

Cronin, Matthew A., Steven C. Amstrup, and Gerald W. Garner. "Interspecific and intraspecific mitochondrial DNA variation in North American bears (*Ursus*)." *Canadian Journal of Zoology* 69 (1991): 2985–2992.

Derocher, Andrew, Nicholas Lunn, and Ian Stirling. "Polar bears in a warming climate." *Integrated and Comparative Biology* 44 (2004): 163–76.

Derocher, A. E., and Ø. Wigg. "Infanticide and cannibalism of juvenile polar bears (*Ursus maritimus*) in Svalbard." *Arctic* 52, no. 3 (1999): 307–10.

Dyck, M. G., W. Soon, R. K. Baydack, D. R. Legates, S. Baliunas, T. F. Ball, and L. O. Hancock. "Polar bears of western Hudson Bay and climate change: Are warming spring air temperatures the 'ultimate' survival control factor?" *Ecological Complexity* 4 (2007): 73–84.

Fischbach, A. S., S. C. Amstrup, and D. C. Douglas. "Landward and eastward shift of Alaskan polar bear denning associated with recent sea ice changes." *Polar Biology* 30 (2007): 1395–1405.

Harington, C. R. "Denning habits of the polar bear (*Ursus maritimus* Phipps)." *Canadian Wildlife Service Report Series* 5 (1968): 1–32.

Harington, C. R. "The evolution of Arctic marine mammals." *Ecological Applications* 18 (2008): S23–S40.

Heaton, T. H., S. L. Talbot, and G. F. Shields. "An ice age refugium for large mammals in the Alexander Archipelago, southeastern Alaska." *Quaternary Research* 46 (1996): 182–92.

Herrero, Stephen, and Susan Fleck. "Injury to people inflicted by black, grizzly or polar bears: Recent trends and new insights." *Bears: Their Biology and Management*, vol. 8: *A Selection of Papers from the Eighth International Conference on Bear Research and Management, Victoria, British Columbia, Canada, February 1989* (1990), pp. 25–32.

Kelly, Brendan P. "Ringed Seal, *Phoca hispida.*" In *Selected Marine Mammals of Alaska: Species Accounts with Research and Management Recommendations,* edited by Jack W. Lentfer, 57–76. Washington, DC: Marine Mammal Commission, 1988.

Larsen, Thor S., and Ian Stirling. *The Agreement on the Conservation of Polar Bears — Its History and Future.* Tromsø: Norwegian Polar Institute, 2009. 16 pages.

Lunn, Nick J., Ian Stirling, and Dennis Andriashek. "Selection of maternity dens by female polar bears in western Hudson Bay, Canada and the effects of human disturbance." *Polar Biology* 27 (2004): 350–56.

Marz, Stacey, and Monica Medina. *On Thin Ice: The Precarious State of Arctic Marine Mammals in the United States Due to Global Warming.* Yarmouth Port, MA: International Fund for Animal Welfare, 2007. 39 pages.

Monnett, Charles, and Jeffrey S. Gleason. "Observations of mortality associated with extended open-water swimming by polar bears in the Alaskan Beaufort Sea." *Polar Biology* 29 (2006): 681–87.

Moore, Sue E., and Henry P. Huntington. "Arctic marine mammals and climate change: Impacts and resilience." *Ecological Applications* 18, no. 2 (2008): S157–S165.

Mulvaney, Kieran. "Arctic Blast." *Washington Post Magazine,* September 13, 2009, pp. 22–29, 32.

O'Neill, Saffron J., Tim J. Osborn, Mike Hulme, Irene Lorenzoni, and Andrew R. Watkinson. "Using expert knowledge to assess uncertainties in future polar bear populations under climate change." *Journal of Applied Ecology* 45 (2008): 1649–59.

Regehr, Eric V., Nicholas J. Lunn, Steven C. Amstrup, and Ian Stirling. "Effects of earlier sea ice breakup on survival and population size of polar bears in western Hudson Bay." *Journal of Wildlife Management* 71, no. 8 (2007): 2673–83.

Richardson, E., I. Stirling, and B. Kochyubajda. "The effects of forest fires on polar bear maternity denning habitat in western Hudson Bay." *Polar Biology* 30 (2007): 369–78.

Schliebe, Scott, Thomas Evans, Kurt Johnson, Michael Roy, Susanne Miller, Charles Hamilton, Rosa Meehan, and Sonja Jahrsdoerfer. *Range-Wide Status Review of the Polar Bear* (Ursus maritimus). Anchorage, AK: U.S. Fish & Wildlife Service, 2006. 262 pages.

Shnayerson, Michael. "The Edge of Extinction." *Vanity Fair,* May 2008, pp. 246–53, 275–78.

Stirling, Ian. "Polar Bear: *Ursus maritimus.*" In *Encyclopedia of Marine Mammals,* edited by W. F. Perrin, B. Würsig, and J. G. M. Thewissen, 945–47. San Diego: Academic Press, 2002.

Stirling, Ian, and Andrew E. Derocher. "Melting under pressure: The real scoop on climate warming and polar bears." *Wildlife Professional* (Fall 2007): 24–27, 43.

Stirling, Ian, and Andrew E. Derocher. "Possible impacts of climate warming on polar bears." *Arctic* 46, no. 3 (1993): 240–45.

Stirling, Ian, and Claire Parkinson. "Possible effects of climate warming on selected populations of polar bears (*Ursus maritimus*) in the Canadian Arctic." *Arctic* 59, no. 3 (2006): 261–75.

Stirling, Ian, Andrew E. Derocher, William A. Gough, and Karyn Rode. "Response to Dyck et al. (2007) on polar bears and climate change in western Hudson Bay." *Ecological Complexity* 5 (2008): 193–201.

Stirling, Ian, Nicholas J. Lunn, and John Iacozza. "Long-term trends in the population ecology of polar bears in western Hudson Bay in relation to climatic change." *Arctic* 52, no. 3 (1999): 294–306.

Talbot, S. L., and G. F. Shields. "A phylogeny of the bears (Ursidae) inferred from complete sequences of three mitochondrial genes." *Molecular Phylogenetics and Evolution* 5, no. 3 (1996): 567–75.

Taylor, Mitchell, Thor Larsen, and R. E. Schweinsburg. "Observations of intraspecific aggression and cannibalism in polar bears (*Ursus maritimus*)." *Arctic* 38, no. 4 (1985): 303–309.

Thiemann, Gregory W., Andrew E. Derocher, and Ian Stirling. "Polar bear *Ursus maritimus* conservation in Canada: An ecological basis for identifying designatable units." *Oryx* 42, no. 4 (2008): 504–15.

Yu, Li, Qing-wei Li, O. A. Ryder, and Ya-ping Zhang. "Phylogeny of the bears (Ursidae) based on nuclear and mitochondrial genes." *Molecular Phylogenetics and Evolution* 32 (2004): 480–94.

Index